AF516163

An Introduction to
ENVIRONMENT

An Introduction to ENVIRONMENT

Vishwambhar Prasad Sati

RAWAT PUBLICATIONS

Jaipur • New Delhi • Bangalore • Hyderabad • Guwahati

ISBN 978-81-316-0497-7 (HB)
ISBN 978-81-316-0498-4 (PB)

© Author, 2012

No part of this book may be reproduced or transmitted in any form or by any means, electronic or mechanical, including photocopying, recording or by any information storage and retrieval system, without permission in writing from the publishers.

Published by
Prem Rawat for **Rawat Publications**
Satyam Apts, Sector 3, Jawahar Nagar, Jaipur 302 004 (India)
Phone: 0141 265 1748 / 7006 Fax: 0141 265 1748
E-mail: info@rawatbooks.com
Website: www.rawatbooks.com

New Delhi Office
4858/24, Ansari Road, Daryaganj, New Delhi 110 002
Phone: 011 2326 3290

Also at *Bangalore, Hyderabad* and *Guwahati*

Typeset by Rawat Computers, Jaipur
Printed at Nice Printing Press, New Delhi

Contents

List of Tables, Figures and Boxes

Tables

Figures

Boxes

Preface

Environmental studies have recently possessed a status of core discipline among all subjects taught at different levels – primary, secondary and higher. In the wake of high level of environmental degradation and thus, worldwide awareness towards conserving environment, environmental studies got a prime concern to be taught to the students at all given levels. Considering the growing significance of environmental studies, I felt an urgent need to compose a treatise on it according to the course contents of higher classes.

Today, biotic and abiotic means are struggling for their existence because of environmental degradation on the entire earth. Tremendous urbanization, industrialization, population growth and large-scale exploitation of natural resources without framing and implementing any rationale measures of conservation have together led the way to high level temporal and spatial changes in the environmental conditions – climate change, global warming and cooling in particular. Summers are too hot, winters are too cold, also extreme floods and extreme droughts put the constraints for the survival of living beings. Environmental hazards – natural and man-made – on the other hand, are more intensified and extreme.

Currently, two thoughts have arisen from this situation worldwide. A group of people advocate that the phenomenon of climate change is natural. Temporal and spatial variations in the climatic conditions are the natural phenomenon and have also occurred during the past, continue at present and will be continued in the future. The earth came under severe cooling for four consecutive times during the past geological timescale and these periods are nomenclature of various ice ages. Similarly, there were extreme warmer periods at the same time. Another group of people argue that changes in climatic conditions are though natural yet, the way conditions are changing is unique and did not take place earlier. Keeping the opinion of these groups aside, it is undoubtedly accepted that the entire world is under the severe changes of climatic conditions. These all take place due to man-induced activities – exploitation of natural resources, extreme carbon emission, and interfering with the nature for enhancing infrastructural facilities. Mining, quarrying, blasting, construction of roads, railways, large-scale felling of trees, construction of dams and settlements, etc., further accentuate the intensity of environmental degradation.

This book is comprised of 15 chapters. First chapter deals with an introduction to environment – meaning and definition, scope of environment, elements of environment, various types of environment, structure of environment, factors threatening the environment, efforts to protect environment, international treaties, green parties, need for public awareness, institutions in environment, and the thinkers in the area of environment. Second chapter studies natural resources – definition and classification of natural resources and reserves; renewable resources – forests, solar energy, wind energy, hydroelectricity and biofuel; and non-renewable resources – coal, petroleum and resources availability. Conservation of natural resources is studied in Chapter 3. It includes history of conservation, methods of conservation, current conservation issues, biodiversity conservation, forest conservation, soil conservation, water conservation and energy conservation. Types of watershed, components of watershed management, land management, water management, biomass management, people's participation and sustainability are discussed in Chapter 4. Chapter 5 denotes an important content of the book, i.e., ecosystem. Biomes and biosphere, national parks and preserves, ecosystem, energy and nutrients, imbalances, population and communities, diversity, habitat and niche, population growth rate, community interaction, competition, predation, parasitism, co-evolution and succession, and climax community are the major contents within this chapter. Ecosystems – classifications, functions of ecosystems, trophic level, energy flow ecosystem management is dealt with in Chapter 6. Chapter 7 studies biodiversity, which includes global biodiversity crisis, human impacts and preserving diversity. Environmental pollutions – air pollution, water pollution, noise pollution, soil pollution and thermal pollution – are the contents of Chapter 8. Chapter 9 deals with environmental hazards – introduction, types of environmental hazards, natural hazards, terrestrial and atmospheric, man-made, natural disaster reduction and management, and risk and risk reduction. Environmental degradation is discussed in Chapter 10. It includes types of environmental degradation, soil erosion, soil salinisation, deforestation, bush encroachment, loss of biodiversity, causes of environmental degradation and solution to environmental degradation. Chapter 11 studies population explosion and consequences – growth of population, consequences of population explosion in Indian context, birth and death rates, trends of poverty and its environmental effects, environmental challenges and policy implications. Chapter 12 deals in global warming and climate change. It further discusses the global warming and how it is measured, salient features of global warming, basics of global warming and consequences of global warming. Environmental and social issues are discussed in Chapter 13. The main contents in this chapter are environmental issues, issues in environmental policies, towards sustainable development, major problems related to energy, resettlements and rehabilitation of people, environmental ethics, wasteland reclamation, issues involved in enforcement of environmental legislations and environmental education. Chapter 14 deals with the conservation movements – history of conservation ideas, leaders of conservation movement, major conservation movements and climate change conventions. Last chapter looks at India's major environmental regulations such as the Forest Conservation Act, the Environmental Protection Act, the Air Prevention and Control of Pollution Act, the Water Prevention and Control Act, the Wildlife Protection Act, the Biological Diversity Act, and the Electricity Act.

This work is the result of my continuous and everlasting study on the environment. Being a student of geography, I studied physical geography, environmental geography, biogeography, and social and economic geography. The aspects and contents of the above-cited subject matters are well included in this book. Literature related to the environment is taken as references from the books, journals, encyclopedias and internet. I sincerely acknowledge the contributions I took from these means and dedicate this work to those who are working with environmental and social issues for the well-being of the society and its people.

Vishwambhar Prasad Sati

Acronyms and Abbreviations

AEA	Atomic Energy Act
AIS&LUS	All India Soil and Land Use Surveys
BNHS	Bombay Natural History Society
BSI	Botanical Survey of India
BVIEER	Bharati Vidyapeeth Institute of Environment Education and Research
BDA	Biological Diversity Act
CDC	Control and Prevention
CEE	Centre for Environment Education
CERCLA	Comprehensive Environmental Response, Compensation and Liability Act
CFCs	Chlorofluorocarbons
CITES	Convention on International Trade in Endangered Species
CSE	Centre for Science and Environment
CA	Compensatory Afforestation
CAMPA	Compensatory Afforestation Fund Management and Planning Authority

CCC	Copenhagen Climate Convention
CBWPC	Central Board of Water Pollution Control
CDC	Centres for Disease Control and Prevention
CDM	National Clean Development Mechanism
CETPs	Common Effluent Treatment Plants
CNG	Compressed Natural Gas
CREP	Charter of Corporate Responsibility for Environmental Protection
CRZ	Coastal Regulation Zone
CWW	Chief Wildlife Warden
DDT	Dichlorodiphenyltrichloroethane
DNA	Deoxyribonucleic Acid
ENSO	El Nino/southern oscillation
EOLSS	Encyclopedia of Life Support Systems
EOS	Earth Observing System
EPA	Environmental Protection Agency
FP	Forest Proposal
GCP	Global Change Programme
GIS	Geographical Information Systems
GNP	Gross National Product
HDGC	Human Dimension of Global Change
HSWA	Hazardous and Solid Waste Amendments
MAB	Man and the Biosphere Programme
MBIs	Market Based Instruments
MOP	Ministry of Power
NASA	National Aeronautics and Space Administration
NGO	Non-Governmental Organization
NMFS	National Marine Fisheries Service
NPL	National Priority List
NBWL	National Board for Wildlife

NEERI	National Environmental Engineering and Research Institute
NOC	No Objection Certificate
NRA	Natural Resource Accounting
ODS	Ozone Depleting Substances
OSHA	Occupational Safety and Health Administration
PCBs	polychlorinated biphenyls
PIL	Public Interest Litigation
RCRA	Resource Conservation and Recovery Act
SPCB	State Pollution Control Board
SPM	Suspended Particle Matter
SACON	Salim Ali Centre for Ornithology and Natural History
SARA	Superfund Amendments and Reauthorization Act
SDWA	Safe Drinking Water Act
SPCB	State Pollution Control Board
TED	Turtle Excluder Device
TSCA	Toxic Substances Control Act
UNEP	United Nations Environmental Programme
UNFCCC	United Nations Framework Convention on Climate Change
UKSN	Uttarakhand Seva Nidhi
UN	United Nations
UNCED	United Nations Conference on Environment and Development
UNESCO	United Nations Educational, Scientific and Cultural Organization
USGS	United States Geological Survey
USSR	Union of Soviet Socialist Republics
UV	Ultraviolet

VOCs	Volatile Organic Chemicals
WCED	World Commission on Environment and Development
WCP	World Climate Programme
WHO	World Health Organization
WMC	Waste Minimization Circles
WPCPA	Water Prevention and Control of Pollution Act
WII	Wildlife Institute of India
WWF	World Wide Fund
ZSI	Zoological Survey of India

CHAPTER 1

An Introduction to Environment

Environment is a comprehensive term which refers to our surroundings and includes all components – living and non-living (lithosphere, atmosphere and hydrosphere) that are present on the earth. Further, environment is sub-divided into natural and social environment. While the natural environment denotes land, water and air (gases) – the life supporting components, social environment refers to man, society and their interaction with the natural environment. The relationship between the two is called ecology. In other words, ecology studies relationship between biotic and abiotic means. A study of interrelationship between living and non-living being – how living beings interact with non-living beings and how non-living beings get impacted with living beings – is called ecosystem. The study of all components of environment – natural and social – is known as environmental studies. It is an interdisciplinary science as it contains various branches of studies like earth sciences, physical sciences, agricultural sciences and social sciences. The subject matter of environmental sciences includes the study of land, water and air (natural environment) and population, settlements, economy, social issues and man-environment relationship (social environment). During the recent past, due to increase in human population and thus high level of economic activities, the natural environment has degraded at a large scale. The environmental and social issues ranging from high level of environmental pollution, exploitation of natural resources and environmental degradation to health hazards manifested a way for regulation of laws for conserving environment. The world communities – governments, administrators, policy institutions, academia and local people – are all concentrating on these issues.

A basic principle known as the 'Law of the Minimum' was propounded by a German chemist Justus Von Liebig in 1840, which says that population cannot grow indefinitely as it is the fact that biotic and abiotic factors, singly or in combination, ultimately limit the size that any population may attain. This size limit, known as a population's carrying capacity, occurs when needed resources – food, breeding sites and water – are in short supply. For example, the amount of nutrients in soil influences the amount of crop grown in a farm. Population size and distribution may also be affected, either directly or indirectly, by the way species in an ecosystem interact with one another. Typically, the species that coexist in ecosystems have evolved together for many generations. They have established balanced interactions with each other that enable whole population of an area to remain relatively stable. Occasionally, however, natural or human-made disruptions occur that have unforeseen consequences to people in an ecosystem. To better understand the impact of natural and human disruptions on the earth, in 1991 the National Aeronautics and Space Administration (NASA) began to send artificial satellites into space study global change. NASA's undertaking, called Earth Science Enterprise, is part of an international effort linking numerous satellites into a single Earth Observing System (EOS). EOS collects information about the interactions occurring in the atmosphere, on land, and in the oceans, and these data help scientists and lawmakers to make sound environmental policy decisions.

Meaning and Definitions

All the external factors influencing the life and activities of people, plants and animals is called environment. Every issue that affects an organism is studied in environmental science. It is essentially a multidisciplinary approach that brings about an appreciation of our natural world and human impacts on its integrity. It is an applied science as it seeks practical answers in making human civilization sustainable on the earth's finite resources. Its components include geography biology, geology, chemistry, physics, engineering, sociology, health, anthropology, economics, statistics, computers and philosophy. Literally, environment means the surrounding external conditions influencing development or growth of people, animal or plants, living or working conditions, etc. This involves three questions – what, by what and where is it surrounded? The answer to this question is living objects in general and man in particular and the physical attributes, which become environment. In fact, the concern of all education is the environment of man. However, man cannot exist or be understood in isolation from the other forms of life. Hence, environment refers to the sum total of condition, which surrounds point in space and time. The scope of the term 'environment' has been changing and widening with the passage of time. Man has extended his environment through his social, economic and political functions. It is in nature that physical components of the planet earth, viz., land, air, water, etc., support and affect life in the biosphere. According to A. Goudie, "environment is the representative of physical components of the earth wherein man is an important factor affecting the environment". Some important definitions of environment are as under:

(a) *Boring:* "A person's environment consists of the sum total of the stimulation which he receives from his conception until his death." It can therefore be said that environment comprises various types of forces such as physical, intellectual, economic, political, cultural, social, moral and emotional. It is the assemblage of all the external forces, influences and conditions, which affect the life, nature, behaviour and the growth, development and maturation of living organisms.
(b) *Douglas and Holland:* "The term 'environment' is used to describe, in the aggregate, all the external forces, influences and conditions, which affect the life, nature, behaviour and the growth, development and maturity of living organisms."
(c) *International Epidemiological Association (2001):* "All that which is external to the human host is an environment. It can be divided into physical, biological, social and cultural, any or all of which can influence health status of populations." According to

this definition, the environment would include anything that is not genetic, although it could be argued that even genes are influenced by the environment in the short or long term.

Figure 1.1 Definition of Environment

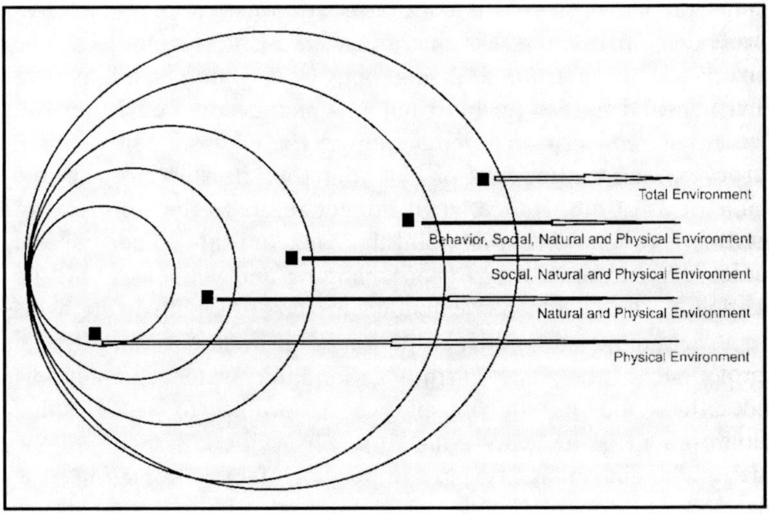

Source: Adopted from Smith, Carvalan and Kjellstorm (1991).

Keeping the above-cited definitions in view, we can conclude that environment refers to all the substances that occur in our surrounding including land, water and air (life supporting layers), and living organisms (animal and plant kingdom). It also studies the relationship and interrelationship between living and non-living organisms.

Scope of Environment

Our surrounding is originally a natural landscape – a forest, a river, a mountain, a desert, or a combination of all these elements. Most of the landscapes have been heavily modified by human beings in villages, towns or cities. But even the people living in cities get their food supply from the surrounding villages. Living organism is fully dependent on natural landscapes such as forests, grasslands, rivers and seashores for resources such as water for agriculture, fuel wood, fodder and fish. Thus, people's daily lives are linked with the surroundings which inevitably affect them. For example, water is used for drinking and other day-to-day activities, air to breathe and resources for food. Everything around us forms environment and lives depend on keeping its vital systems as intact as possible. Dependence on nature is so great that life cannot exist without protecting the earth's environmental resources. Thus, we call environment as 'Mother Nature' and most traditional societies have learned that respecting nature is vital for their livelihoods. This has led to many cultural practices that helped traditional societies to protect and preserve their natural resources. Respect for nature and all living creatures is not new to India. All our traditions are based on the values that conserve the environment. Emperor Ashoka's edict in 4th century BC proclaimed that all forms of life are important for our well-being.

Over the past decades, innovations in science and technology provided the way for economic growth through using high-yield seeds and fertilizers in the field of agriculture, improved varieties of animals, irrigation facilities through construction of large-scale dams, industrial development and overutilization of resources to feed the growing population. This has led to extreme environmental degradation. The industrial development and intensive agriculture that provides the goods for increasingly consumer-oriented society uses large amounts of natural resources – water, minerals, petroleum products and wood. Non-renewable resources such as minerals and oil are being extracted from nature in an unsustainable manner and will be exhausted in the future if we continue to do so. Sustainable use of renewable resources in lieu of non-renewable resources will conserve the nature and protect the life on the earth. But renewable resources will also be depleted if used faster than nature can replace them. For example, if the removal of timber and firewood from a forest is faster than the re-growth and regeneration of trees, it cannot replenish the supply. Loss of forest cover not only depletes the forest of its resources, such as timber and other non-wood products, but affects our water resources too. Deforestation leads to floods in the monsoon and dries rivers once the rains are over. Thus, it is obvious that the environment has multiple impacts on the living creatures as it sustains life of humans and animals both. Thus, sustainable utilization or development of natural resources is the need of the hour that can maintain our future and also the next generation.

Box 1.1 Segments of Environment

1. *Atmosphere:* The atmosphere is composition of gases, dust particles and water vapours that surrounds the earth. It fulfils the following functions:
   - It sustains life on the earth.
   - It saves us from the hostile environment of outer space.
   - It absorbs most of the cosmic rays from outer space and a major portion of the electromagnetic radiation from the sun.
   - It transmits only here ultraviolet, visible, near infrared radiation (300 to 2,500 nm) and radio waves. The atmosphere is composed of nitrogen, oxygen, argon, carbon dioxide and trace gases.
2. *Hydrosphere:* Hydrosphere comprises all types of water reservoirs – oceans, seas, lakes, rivers, streams, polar icecaps, glaciers and ground water. It has the following characteristics:
   - About 72 per cent of the earth's surface is covered by hydrosphere.
   - Out of this, oceans cover 97 per cent area.
   - About 2 per cent of the water resources are locked in the polar icecaps and glaciers.
   - Only about 1 per cent is available as fresh surface water rivers, lakes streams, and ground water, fit to be used for human consumption and other uses.
3. *Lithosphere:* Lithosphere is the outer surface of the earth. It consists of minerals occurring in the earth's crusts and the soil, e.g., minerals, organic matter, air and water.
4. *Biosphere:* Biosphere comprises living organisms and their interactions with environment – atmosphere, hydrosphere and lithosphere.

The biotic environment can be divided into floral and faunal environments. All the organisms work to form their social groups and organizations at several levels. Thus, the social environment is formed. In this social environment the organisms work to derive matter from the physical environment for their sustenance and development. This process gives birth to economic environment. Man claims to be most skilled and civilized of all the organisms. This is the reason why his social organization is most systematic.

Box 1.2 Various Types and Elements of Environment

Environment is constituted by the interacting systems of physical, biological and cultural elements, interrelated in various ways, individually as well as collectively. According to Kurt Lewin, environment is of three types which influence the personality of an individual as under:

(i) *Physical Elements:* Physical elements include space, landforms, water bodies, climate, soils, rocks and minerals. They determine the variable character of the human habitat, its opportunities as well as limitations. Physical environment refers to geographical climate and weather or physical conditions wherein an individual lives. The human races are greatly influenced by the climate. Some examples are as under:

- In the cold countries such as European countries, North America, Siberian Asia and North-East Asia, the people are of white colour. Likewise, in South Asian and African countries (hot countries), the people are of dark complexion.
- The physique of an individual depends on climatic conditions as the individual tries to adjust in his physical environment.
- The human working efficiency also depends on the climatic conditions.

(ii) *Biological Elements:* Various elements – plants, animals, microorganisms and men – constitute the biosphere. Although, physical and social environments are common to the individual in a specific situation yet, every individual has his own psychological environment, in which he lives. Lewin has used the term 'life space' for explaining psychological environment. The psychological environment enables us to

Cont'd...

...Cont'd

understand the personality of an individual. Both the person and his goal form psychological environment. If a person is unable to overcome the barriers, he can either get frustrated or may change his goal for a new psychological environment. By adopting this mechanism, the individual is helped in his adjustment to the environment.

(iii) *Cultural and Social Elements:* Cultural elements – economic, social and political – are essentially man-made features, which make cultural milieu. Social environment includes an individual's social, economic and political conditions in which he lives. The moral, cultural and emotional forces influence the life and nature of individual behaviour. Society may be classified into two categories as under:

- An open society, which is very conducive to individual development.
- A closed society, which is not very supportive to development.

Man and Environment

There are three aspects of man, viz., physical, social and economic, which function in the biotic environment. The 'Physical Man' is one of the organism's populations or biological community. He is in need of basic elements of the physical environment like habitat (space), air, water and food. Besides, like other biological populations, he releases wastes into the ecosystem. The 'Social Man' performs the following functions: establishing social institutions, forming social organizations, formulating laws, principles and policies and taking steps to safeguard his existence, interest and social welfare. The 'Economic Man' derives and utilizes resources from the physical and biotic environment with his skills and technologies. The economic function makes the man an environment process as he transports matter and energy from one component of the ecosystem to the other. There may be two situations: his exploitative functions may be in harmony with the natural environment. Such, functions do not necessarily involve change in the working of the ecosystem and may exceed the critical limit.

Consequently, the equilibrium of the environment is disturbed and a great number of environment and ecological problems crop up. These are detrimental to man himself besides to whole population of human species in a given ecosystem.

Figure 1.2 Structure of Environment

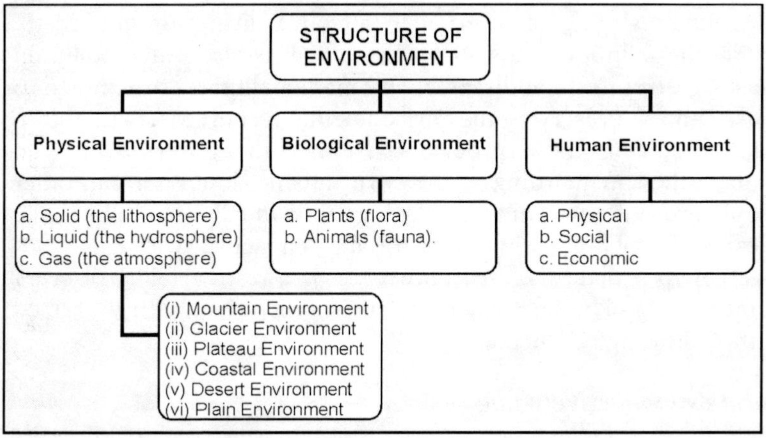

Importance of Environmental Studies

Environment is the combination and integration of multidisciplinary subjects. It includes both natural and social sciences. To understand the different aspects of the environment, it is required to understand biology, chemistry, physics, geography, resource management, economics and population issues. Thus, the scope of environmental studies is extremely comprehensive and wide and covers aspects of every major discipline. At present, natural resources, particularly non-renewable resources, are limited in the world. Water, air, soil, minerals, oil, forest and its products, grasslands, oceans, agriculture and agricultural products and livestock are crucial for sustenance of life on the earth. Life cannot exist without these life supporting systems. As the unexpected growth of population is taking place, increasing use of natural resources is spontaneous and consequently, the natural resources are inevitably shrinking. As the natural resources are limited, the earth cannot be expected to sustain this expanding level of utilization of resources. Misuse of resources is further accelerating the shrinkage of resources. Pollution is another major reason for degrading environment. High rate of water pollution has created drinking water scarcity. Material like plastic and large amount of food waste is discarded as garbage. Besides, the manufacturing industries also produce both solid and liquid wastes. Increasing amounts of waste cannot be managed by natural processes. These accumulate in the environment, leading to a variety of diseases and other adverse environmental impacts, seriously affecting living on the earth. While air pollution leads to respiratory diseases, water pollution causes gastrointestinal diseases, and many pollutants are known to cause cancer. Preserving the environmental resources will therefore help in solving these problems. Neither the governments can manage the safeguarding of the environment alone, nor can other people prevent environmental damage individually. It is all people, irrespective of any religion or of any capacity, who can work together to protect the environment. It is a responsibility of every human being. The importance of environment has been described in the following sub-heads:

Nature's Productive Value

Scientists worldwide have made new advances in the field of biotechnology and understood that the world's species contain an incredible and uncountable number of complex chemicals. The raw materials, used for developing new medicines and industrial products, is a storehouse from which thousands of new products in the future can be developed. The flowering plants and insects that form the most species rich groups of living organisms are thus vital for the future development of man. If degradation of species continues, these will become extinct. The government should take necessary action to stop illegal poaching of wild species. Everybody should in this regard help the authorities to save wild species from being extincted. Humans cannot bring them back if they are vanished completely. It will have adverse consequences on the future generations too, if immediate steps to protect environment are not taken. There is a vital need to conserve the forests, wetlands, and the natural areas so that both the present and next generations can be protected. It is not an individual who can stop degradation of environment; it is the community involvement which can better enhance the environment. Wildlife sanctuaries, national parks and other natural areas should be protected carefully. There is a close link between agriculture and the forest, which illustrates its productive value. For crops to be successful, the flowers of fruit trees and vegetables must be pollinated by insects, bats and birds. Their life cycles however frequently require intact forests.

Nature's Recreational Value

The earth has lot of areas where aesthetic and recreational values maintain existence of life. National parks and wildlife sanctuaries are among them as they are relatively undisturbed areas. A true wilderness experience has not only recreational value but is an incredible learning experience. It brings about an understanding of the oneness of nature and the fact that lives are entirely dependent upon the intricate functioning of ecosystems. The beauty of nature encompasses every aspect of both living and non-living parts of the earth. The mighty mountains, the widespread seas, the wonderful forests and the deserts make this world more magnificent, spectacular and beautiful. These natural vistas and their incredible diversity of plant and animal life have led to the development of several philosophies of life. It has also inspired artists to develop visual arts and writers and poets to create their works that could energize our lives. A wilderness experience has exceptional recreational value. This has been described as nature tourism, or wildlife tourism, and is also one aspect of adventure tourism. These recreational facilities not only provide a pleasurable experience but are also intended to create a deep respect and love for nature. They are also key tools in educating people about the fragility of the environment and the need for sustainable lifestyles. In an urban setting, green spaces and gardens are vital to the psychological and physical health of city dwellers. They provide not only an aesthetic and visual appeal but the ability to ensure that each individual is able to access a certain amount of peace and tranquillity. Thus, urban environmental planners must ensure that these facilities are created in growing urban complexes. Another important conservation education facility in an urban setting includes the need to set up well-designed and properly managed zoological parks and aquariums. These have got great value in sensitizing school students to wildlife. Many young people who frequented zoos as young children grow up to love wildlife and become conservationists. In the absence of access to a protected area, a botanical garden or a zoo, one concept that can be developed is to create small nature awareness areas with interpretation facilities at district levels. These areas can be developed to mimic natural ecosystems even though they could be relatively small in size. Such nature trails are invaluable assets for creating conservation education and awareness. They can be developed in a small woodlot, a patch of grassland, a pond ecosystem, or can be situated along an undisturbed river or coastal area. This would bring home to the visitor the importance of protecting our dwindling wilderness areas.

Nature's Option Values

We utilize several goods and services of nature and enjoy their benefits. Every human action of development has reaction on the environment and nature's integrity. During the past, human being has exploited natural resources at a large scale and, consequently, it has adverse impact on the environment. High level of pollution, killing of wildlife and depletion of forest have become severe problems for the future generation. Nature provides us with various options on how we should utilize its goods and services. This is its option value. Resources are being exploited greedily, which is destroying its integrity and long-term values. Sustainable utilization of resources will reduce impacts on the environment. The option value allows using resources sustainably and preserves its goods and services for the future. The environment studies enlighten about the importance of protection and conservation and indiscriminate release of pollution into the environment. At present, a great number of environment issues have grown in size and complexity day by day, threatening the survival of mankind on earth. Environmental studies have become significant for the following reasons:

International Importance of Environmental Issues

It has been well recognized that environmental issues like global warming and ozone depletion, acid rain, marine pollution and biodiversity are not merely national issues but are global issues and hence, must be tackled with international efforts and cooperation.

Consequences of Development

Development, in its wake, gave birth to urbanization, industrial growth, transportation systems, agriculture and housing, etc. However, it has become phased out in the developed world. The North, to cleanse their own environment, has factfully managed to move 'dirty' factories to South. When the West developed, it did so perhaps in ignorance of the environmental impact of its activities. Evidently, such a path is neither practicable nor desirable, even if developing world follows that.

Tremendous Increase in Pollution

World census reflects that one in every seven persons in this planet lives in India. Evidently, 16 per cent of the world's population lives in 2.4 per cent of land area. Consequently, there is a heavy pressure on the natural resources including land. Agricultural experts have recognized problems pertaining to soil like deficiency of micronutrients and organic matter, soil salinity and damage of soil structure.

Need for an Alternative Solution

It is essential, especially for developing countries, to find alternative paths to an alternative goal. We therefore need: (1) A goal, which ultimately is the true goal for an environmentally sound and sustainable development. (2) A goal common to all peoples of our earth. (3) A goal distant from the developing world to the overconsuming wasteful societies of the developed world.

Need to Save Humanity from Extinction

It is incumbent upon us to save the humanity from extinction. In the name of development, severe environmental degradation and depleting of biosphere occurred during the last decades.

Need for Wise Planning of Development

Our survival and sustenance depends on the availability of various resources. Taking resources, processing and finally usage of the product should be synchronized with the ecological cycles in any plan of development. The actions should be planned ecologically for the sustenance of the environment and development.

Misra's Report

Misra (1991) has discerned four basic principles of ecology – holism, ecosystem, succession and conversation. Holism has been considered as the real base of ecology. In hierarchical levels at which interacting units of ecology are discussed, these are presented as follows: individual<population<community< ecosystem< biome<biosphere.

Misra (1991) has also recognized four basic requirements of environmental management: (a) impact of human activities on the environment, (b) value system, (c) plan and design for sustainable development, and (d) environmental education. Keeping in view the of goal of planning for environmentally sustainable development, India contributed greatly to the United Nations Conference on Environment and Development (UNCED), also referred to as "Earth Summit", held at Rio de Janeiro, the capital of Brazil, in 1992.

Factors Threatening the Environment

The problems facing the environment are vast and diverse. Population explosion, global warming, the depletion of the ozone layer and destruction of the world's rain and temperate forests are the major problems that many scientists believe will reach the critical proportions in the coming decades. All of these problems will be directly affected by the size of the human population. The major factors threatening the environment are discussed below:

Population Growth

Human population growth is at the root of virtually all of the world's problems, including environmental problems. Although the growth rate of the world's population has slowed slightly since the 1990s, the world's population increases by about 77 million each year. As the number of people increases, it generates pollution, destroys more habitats, and leads to search for new natural resources. The Population Division of the United Nations predicts that the world's population will increase from 6.23 billion people in 2000 to 9.30 billion in 2050. The UN estimates that the population will stabilize at more than 11 billion in 2200. Other experts predict that it will continue to rise into the foreseeable future, to as many as

19 billion by the year 2200. Although rates of population increase are now much slower in the developed world than in the developing world, it would be a mistake to assume that population growth is primarily a problem of developing countries. In fact, because larger amounts of resources per person are used in developed nations, each individual from the developed world has a much greater environmental impact than does a person from a developing country. Conservation strategies that would not significantly alter lifestyles but would greatly lessen environmental impact are essential in the developed world. In the developing world, meanwhile, the most important factors necessary to lower population growth rates are democracy and social justice. Studies show that population growth rates have fallen in developing areas where better social conditions exist. In these areas, literacy rates have increased and women receive economic status equal to that of men, enabling women to take jobs and own property. In addition, birth control information in these areas is more widely available, and women are free to make their own reproductive decisions.

Global Warming

Like the glass panes in a greenhouse, certain gases in the earth's atmosphere permit the sun's radiation to heat earth. At the same time, these gases retard the escape into space of the infrared energy radiated back out by earth. This process is referred to as the *greenhouse effect*. These gases, primarily carbon dioxide, methane, nitrous oxide, and water vapour, insulate earth's surface, helping to maintain warm temperatures. Without these gases, earth would be a frozen planet with an average temperature of about –18°C instead of a comfortable 15°C. However, the concentration of these gases rises, they trap more heat within the atmosphere, causing worldwide temperatures to rise. Within the last century, the amount of carbon dioxide in the atmosphere has increased dramatically, largely because people burn vast amounts of fossil fuels – coal and petroleum and its derivatives. The average global temperature also has increased by about 0.6°C within the past century. Atmospheric scientists have found that at least half of that temperature increase can be attributed to human activity. They predict that unless drastic action is taken, global temperature will continue to rise by 1.4° to

5.8°C over the next century. Although such an increase may not seem like a great difference, during the last ice age the global temperature was only 2.2°C cooler than it is presently. The consequences of such a modest increase in temperature may be devastating. Already, scientists have detected a 40 per cent reduction in the average thickness of Arctic ice. Other problems that may develop include a rise in sea levels that will completely inundate a number of low-lying island nations and flood many coastal cities, including New York and Miami. Many plant and animal species will probably be driven into extinction; agriculture will be severely disrupted in many regions, and the frequency of severe hurricanes and droughts will likely to increase.

Depletion of the Ozone Layer

The ozone layer, a thin band in the stratosphere, serves to shield earth from the sun's harmful ultraviolet rays. In the 1970s, scientists discovered that chlorofluorocarbons (CFCs) – chemicals used in refrigeration, air-conditioning systems, cleaning solvents and aerosol sprays – may destroy the ozone layer. CFCs release chlorine into the atmosphere, which in turn, breaks down ozone molecules. As chlorine is not affected by its interaction with ozone, each chlorine molecule has the ability to destroy a large amount of ozone for an extended period of time. The consequences of continued depletion of the ozone layer would be severe. Increased ultraviolet radiation would lead to a growing number of skin cancers and cataracts and also reduce the ability of immune systems to respond to infection. Additionally, growth of the world's oceanic plankton – the base of most marine food chains – would decline. Plankton contains photosynthetic organisms that break down carbon dioxide. If plankton population decline, it may lead to increased carbon dioxide levels in the atmosphere and thus to global warming. Recent studies suggest that global warming, in turn, may increase the amount of ozone destroyed. Even if the production of CFCs is immediately banned, the chlorine already released into the atmosphere will continue to destroy the ozone layer for many decades. In 1987, an international pact called the Montreal Protocol on Substances that deplete the ozone layer set specific targets for all nations to achieve in order to reduce emissions of chemicals responsible for the destruction of the ozone layer. Many people hoped that this treaty would cause ozone loss to peak and begin to decline by the year 2000. In fact, in the fall of 2000, the hole in the ozone layer over Antarctica was the largest than recorded. However, in the following year, it was slightly smaller, leading some to believe that the depletion of ozone has stabilized. However, in 2006, the US scientists reported that the ozone loss over Antarctica reached its greatest extent ever. Ozone loss can vary with temperature, and many scientists believe that the 2006 record loss was due to lower-than-normal temperatures. These scientists report that although CFC levels in the atmosphere peaked in 2001, many of these chemicals are long-lasting. They believe that the ozone layer over Antarctica may not fully recover until 2065.

Habitat Destructions and Species Extinction

Plant and animal species are dying out at an unprecedented rate. According to an estimate, from 4,000 to as many as 50,000 species per year become extinct. The leading cause of extinction is habitat destruction, particularly of the world's richest ecosystems – tropical rainforests and coral reefs. If the world's rainforests continue to be cut down at the current rate, they may completely disappear by the year 2030. In addition, if the world's population continues to grow at its present rate and puts even more pressure on these habitats, they might well be destroyed sooner.

Air and Water Pollution

A significant portion of industry and transportation burns fossil fuels, such as gasoline. When these fuels burn, chemicals and particulate matter are released into the atmosphere. Although a vast number of substances contribute to air pollution, the most common air pollutants contain carbon, sulfur and nitrogen. These chemicals interact with one another and also with ultraviolet radiation in sunlight, in dangerous ways. Smog forms when nitrogen oxides react with hydrocarbons in the air to produce aldehydes and ketones. It can cause serious health problems. Acid rain forms when sulfur dioxide and nitrous oxide transform into sulfuric acid and nitric acid in the atmosphere and come back to earth in precipitation. Acid rain has made numerous lakes so acidic that they no longer support fish populations. Acid rain is also responsible for the decline of many forest ecosystems worldwide, including Germany's Black Forest and forests throughout the eastern United States. Estimates suggest that nearly 1.5 billion people worldwide lack safe drinking water and that at least 5 million deaths per year can be attributed to waterborne diseases. Water pollution may come from point sources or non-point sources. Point sources discharge pollutants from specific locations, such as factories, sewage treatment plants and oil tankers. The technology exists to monitor and regulate point sources of pollution, although in some areas this occurs only sporadically. Pollution from non-point sources occurs when rainfall or snowmelt moves over and through the ground. As the runoff moves, it picks up and carries away pollutants, such as pesticides and fertilizers, depositing the pollutants into lakes, rivers, wetlands, coastal waters, and even underground sources of drinking water. Pollution arising from non-point sources accounts for a majority of the contaminants in streams and lakes. With almost 80 per cent of the planet covered by oceans, people have long acted as if those bodies of water could serve as a limitless dumping ground for wastes. However, raw sewage, garbage and oil spills have begun to overwhelm the diluting capabilities of the oceans, and most coastal waters are now polluted, threatening marine wildlife. Beaches around the world close regularly, often because the surrounding waters contain high levels of bacteria from sewage disposal.

Groundwater Depletion and Contamination

Water that collects beneath the ground is called groundwater. Worldwide, groundwater is 40 times more abundant than fresh water in streams and lakes. In the United States, approximately half the drinking water comes from groundwater. Although groundwater is a renewable resource, its reserves replenish relatively slowly. Presently, groundwater in the United States is withdrawn approximately four times faster than it is naturally replaced. The Ogallala Aquifer, a huge underground reservoir stretching under eight states of the Great Plains, is drawn down at rate exceeding 100 times the replacement rate. In India, groundwater withdrawn is doubling every year. Groundwater contamination is another major issue in India. Agricultural practices depending on this source of water need to change within a generation in order to save this groundwater source. In addition to groundwater depletion, scientists also worry about groundwater contamination, which arises from leaking underground storage tanks, poorly designed industrial waste ponds, and seepage from the deep-well injection of hazardous wastes into underground geologic formations. By some estimates, on an average, 25 per cent of usable groundwater is contaminated, and in some areas as much as 75 per cent has harmful substances, making it undrinkable.

Chemical Risks

A number of toxic substances that humans encounter regularly may pose serious health risks. Pesticide residues on vegetable crops, mercury in fish, and many industrially produced chemicals may cause cancer, birth defects, genetic mutations, or death. Many chemicals have been found to mimic estrogen – the hormone that controls the development of the female reproductive system in a large number of animal species. Preliminary results indicate that these chemicals, even in small amounts, may disrupt development and lead to a host of serious problems in both males and females, including infertility, increased mortality of offspring, and behavioural changes such as increased aggression.

Environmental Racism

Studies have shown that not all individuals are equally exposed to pollution. For example, worldwide toxic-waste sites are more prevalent in poorer communities. In the United States the single most important factor in predicting the location of hazardous-waste sites is the ethnic composition of a neighbourhood. Three of the five largest commercial hazardous-waste landfills in America are in predominantly black or Hispanic neighbourhoods, and three out of every five black or Hispanic Americans live in the vicinity of an uncontrolled toxic-waste site. The wealth of a community is not nearly as good a predictor of hazardous-waste locations as the ethnic background of the residents, suggesting that the selection of sites for hazardous-waste disposal involves racism.

Environmental racism takes international forms as well. American companies often continue to produce dangerous, banned chemicals and ship them to developing countries. Additionally, the developed world has shipped large amounts of toxic waste to developing countries for less-than-safe disposal. For instance, experts estimate that 50 to 80 per cent of electronic waste produced in the United States, including computer parts, is shipped to waste sites in developing countries, such as China and India. At a waste site in Giuyu, China, labourers with no protective clothing regularly burn plastics and circuit boards from old computers. They pour acid on electronic parts to extract silver and gold, and they smash cathode-ray tubes from computer monitors to remove lead. These activities so pollute the groundwater beneath the site that drinking water is trucked into the area 29 km. away from a town.

Energy Production

The limited supply of fossil fuels, coupled with their contributions to global warming, air pollution and acid rain, makes it clear that alternative forms of energy will be needed to fuel industrial production and transportation. A number of energy alternatives are available, but many of these are unlikely to replace fossil fuels in the foreseeable future because they cost more, produce less energy than fossil fuels, or pose safety risks. A handful of countries produce a portion of their electricity using nuclear energy. But many people oppose it because even a small accident can cause massive devastation. The 1986 accident at the Chernobyl nuclear power plant in the Ukraine scattered radioactive contamination over a large part of Europe. Approximately 2,00,000 people were evacuated, and human health has been seriously affected. Studies in 1999 found that the rate of thyroid cancer in young Ukrainian children was ten times higher than was the norm prior to the accident. One reasonable solution combines conservation strategies with the increased use of solar energy. The price of solar energy relative to traditional fuels has steadily dropped, and if environmental concerns were factored into the cost, solar power would already be significantly cheaper.

Efforts to Protect the Environment

Most scientists agree that if pollution and other environmental deterrents continue at their present rates, the result will be irreversible damage to the ecological cycles and balances in nature upon which all life depends. Scientists warn that fundamental, and perhaps drastic, changes in human behaviour will be required to avert an ecological crisis. To safeguard the healthful environment that is essential to life, humans must learn that earth does not have infinite resources. Earth's limited resources must be conserved and, where possible, reused. Furthermore, humans must devise new strategies that connect environmental progress with economic growth. The future growth of developing nations depends upon the development of sustainable conservation methods that protect the environment while also meeting the basic needs of citizens. Many nations have acted to control or reduce environmental problems. For example, Great Britain has largely succeeded in cleaning up the waters of the Thames and other rivers, and London no longer suffers the heavy smog caused by industrial pollutants. Japan has some of the world's strictest standards for the control of water and air pollutions. In Canada, the Department of Commerce has developed comprehensive programmes covering environmental contaminants. In the United States, the Environmental Protection Agency (EPA) was established in 1970 to protect the nation's natural resources. In addition, the US Congress has provided governmental agencies with legislation designed to protect the environment. Many US states have also established environmental protection agencies. Citizen groups, such as the Sierra Club and the National Audubon Society, educate the public, support environment-friendly legislation, and help assure that federal and state laws are enforced by pointing out violations. During the late 1960s and early 1970s nations began to work together to develop worldwide approaches for monitoring and restricting global pollution. The first major international conference on environmental issues was held in Stockholm, Sweden, in 1972 and was sponsored by the United Nations. This meeting, at which the United States took a leading role, was controversial because many developing countries were fearful that a focus on environmental protection was a means for the developed world to keep the undeveloped world in an economically subservient position. The most important outcome of the conference was the creation of the United Nations Environmental Programme (UNEP). UNEP was designed to be "the environmental conscience of the United Nations," and, in an attempt to allay fears of the developing world, it became the first UN agency to be headquartered in a developing country, with offices in Nairobi, Kenya. In addition to attempting to achieve scientific consensus about major environmental issues, a major focus for UNEP has been the study of ways to encourage *sustainable development* – increasing standards of living without destroying the environment.

International Treaties

Dozens of international agreements have been reached in recent decades in an effort to improve the world's environmental status. In 1975 the Convention on International Trade in Endangered Species (CITES) of Wild Fauna and Flora went into effect to reduce commerce in animals and plants on the edge of extinction. In 1982 the International Whaling Commission agreed to a moratorium on all commercial whaling. Perhaps the most important international agreement was the 1987 Montreal Protocol on Substances that deplete the ozone layer. For the first time, an international pact set specific targets for reducing emissions of chemicals responsible for the destruction of earth's ozone layer. The international community again came together in 1989 to form the Basel Convention on the Control of Trans-boundary Movements of Hazardous Wastes and Their Disposal – a treaty that limits the movement of hazardous wastes between countries. In 1992, the UN Conference on Environment and Development was held in Rio de Janeiro, Brazil. Popularly known as the Earth Summit, this was the largest gathering of world leaders in history. The conference resulted in the signing of two major treaties. The first was an agreement for nations to voluntarily reduce emission of gases leading to global warming, and the second was a pact on biodiversity requiring countries to develop plans to protect endangered species and habitats. At the insistence of the United States, however, the final version of the global warming treaty was dramatically scaled back. The United States was also one of the very few countries that refused to sign the biodiversity treaty. Its representatives objected to a part of the treaty that specified that money to come from the use of natural resources from protected ecosystems, such as rainforests, should be shared equally between the source country and the agency or institution removing the materials. The 1992 agreement on global warming limited each industrialized nation to emissions in the year 2000 that were equal to or below 1990 emissions. However, these limits were voluntary and with no enforcement provisions included in the agreement, it became clear by 1997 that these goals would never be met. At a follow-up conference in Kyoto, Japan, representatives from 160 countries signed the Kyoto Protocol. This agreement called for industrialized nations to reduce emissions to an average of about 5 per cent below 1990 emission levels and to reach this goal between the years 2008 and 2012. For this accord to become international law, however, it had to be ratified by at least 55 countries. The United States has refused to ratify the accord, but Japan and the 15 countries that make up the European Union have ratified it. Even if the Kyoto Protocol does become international law, however, scientists expect that its emission requirements are too minimal to be effective. Some experts predict that a 60 per cent reduction in emissions will be necessary to stabilize the world's climate. In 2002 delegates from nearly 200 countries convened at the World Summit on Sustainable Development in Johannesburg, South Africa, to establish new sustainable development goals for the 21st century. They also negotiated to strengthen commitments from the governments of developed nations to provide aid for sustainable development. Among the outcomes, the 2002 summit created an action plan that called on nations to reduce by half the proportion of people who lack sanitation by 2015, to minimize health and environmental problems caused by chemical pollution by 2020, and to reduce significantly the number of endangered species by 2010.

Green Parties

A desire for environmental change led to the creation of various political parties around the world whose emphasis was largely on environmental protection. The first of these organizations, collectively known as *green parties,* was the Values Party in New Zealand, created in 1972. In 1993, 23 green parties from eastern and western Europe came together to form the European Federation of Green Parties. They hoped that together they would have the leverage necessary to demand that environmental issues such as pollution control, population growth and sustainable development be more fully addressed by various national governments and international bodies. By far the most successful green party has been Bundnis 90/Die Grunen of Germany. In 1998 Bundnis 90/Die Grunen formed a coalition with the newly elected Social Democratic Party of German Chancellor Gerhard Schroder, marking the first time that the green party had entered Germany's national government. Green parties have developed in almost all countries that have open elections, but they have had the largest impact in those nations where proportional representation within a parliamentary system occurs. Thus, the green parties have not played a significant role in American politics. However, some experts believe that in the disputed presidential election of 2000, the votes received by Green Party candidate Ralph Nader, split the vote so that George W. Bush was able to win enough electoral votes to capture the presidency.

Future Prospects

Global environmental collapse is not inevitable. But, the developed world must work with the developing world to ensure that new industrialized economies do not add to the world's environmental problems. Politicians must think of sustainable development rather than economic expansion. Conservation strategies have to become more widely accepted, and people must learn that energy use can be dramatically diminished without sacrificing comfort. In short, with the technology that currently exists, the years of global environmental mistreatment can begin to be reversed.

Need for Public Awareness

The earth's natural resources are dwindling and the environment is being increasingly degraded by human activities. It is evident that something needs to be done. Conserving environment is merely not a responsibility of the government; it is a responsibility of the societies. In the prevention of environmental degradation, people should take part and it must be a part of daily activities. To prevent ill-effects on environment by actions is economically more viable than cleaning up the environment once it is damaged. As an individual, we can play a major role in environment management.

We can reduce wasting natural resources. This can only be made possible through mass public awareness. Mass media such as newspapers, radio and television, strongly influence public opinion. If each of us feels strongly about the environment, the press and media will add to our efforts. Politicians in a democracy always respond positively to a strong publicly supported movement. Thus, if we join an NGO that supports conservation, politicians will make green policies. We are living on spaceship earth with a limited supply of resources. Each of us is responsible for spreading this message to as many people as possible. It is essential to make the public aware of the formidable consequences of the environmental degradation, if not retorted and reformative measures undertaken would result in the extinction of life. We are facing various environmental challenges. It is essential to get the country acquainted with these challenges so that their acts may be eco-friendly. Some of these challenges are as under:

Growing Population

A population of over thousands of millions is growing at 2.11 per cent every year. Over 17 million people are added each year. It puts considerable pressure on its natural resources and reduces the gains of development. Hence, the greatest challenge before us is to limit the population growth. Although population control does automatically lead to development, yet the development leads to a decrease in population growth rates. For this development of the women is essential.

Poverty

India has often been described a rich land with poor people. The poverty and environmental degradation have a nexus between them. The vast majority of the people are directly dependent on the natural resources of the country for their basic needs of food, fuel, shelter and fodder. About 40 per cent people are still below the poverty line. Environmental degradation has adversely affected the poor who depend upon the resources of their immediate surroundings. Thus, the challenge of poverty and the challenge of environment degradation are two facets of the same challenge. The population growth is essentially a function of poverty, because to the very poor, every child is an earner and helper, and global concerns have little relevance for him/her.

Agricultural Growth

The people must be acquainted with the methods to sustain and increase agricultural growth without damaging the environment. High yielding varieties have caused soil salinity and damage to physical structure of soil. A rational method of adopting sustainable agriculture is inevitable.

Need for Sustainable Use of Groundwater

It is essential to rationalize the use of groundwater. Factors like community wastes, industrial effluents and chemical fertilizers and pesticides have polluted surface water and affected quality of the groundwater. It is essential to restore the water quality of rivers and other water bodies such as lakes, which is an important challenge. Also, there must be suitable strategies for conservation of water, provision of safe drinking water and keeping water bodies clean.

Development and Forests

Forests serve catchments for the rivers. With increasing demand of water, plans to harness the mighty rivers through large irrigation projects were made. Certainly, these would submerge forests, displace local people, and damage flora and fauna. As such, the dams on the river Narmada, Bhagirathi and elsewhere have become areas of political and scientific debates. Forests in India have been shrinking for several centuries owing to pressures of agriculture and for other uses. Vast areas were once green, stand today as wastelands. These areas are to be brought back under vegetative cover. The tribal communities inhabiting forests respect the trees and birds and animals that give them sustenance. We must recognize the role of these people in restoring and conserving forests. The modern knowledge and skills of the forest department should be integrated with the traditional knowledge and experience of the local communities. The strategies for the joint management of forests should be evolved in a well-planned way.

Degradation of Land

At present, out of the total 329 mha of land, only 266 mha possess any potential for production. Of this, 143 mha is agricultural land and nearly 85 mha suffer from varying degrees of soil degradation. Of the remaining 123 mha, 40 mha is completely unproductive, whereas the 83 mha is classified as forest land, of which over half is denuded to various degrees. Nearly 406 million heads of livestock have to be supported on 13 mha, or less than 4 per cent of the land classified as pasture land, most of which is overgrazed. Thus, out of 226 mha, about 175 mha or 66 per cent is degraded to varying degrees. Water and wind erosions cause further degradation of almost 150 mha, which can be avoided.

Reorientation of Institutions

The people should be aroused to orient institutions, attitudes and infrastructures, to suit conditions and needs of today. The change has to be brought in keeping in view India's traditions for resource use, management and education. Change should be brought in education, in attitudes, in administrative procedures and in institutions because it affects the way people view technology resources and development.

Reduction of Genetic Diversity

Proper measures to conserve genetic diversity need to be taken. At present, most wild genetic stocks have been disappearing from nature. Wildlife, including the Asiatic lions, are facing danger of loss of genetic diversity. The protected areas network like sanctuaries, national parks and biosphere reserves are isolating populations. So, they are decreasing changes of one group breeding with another. Remedial steps need to be taken to check decreasing genetic diversity.

Adverse Consequences of Urbanization

Nearly 30 per cent Indians live in urban areas (2011). Urbanization and industrialization have given birth to a great number of environmental problems that need urgent attention. Over 30 per cent of urban Indians live in slums. Out of India's 3,245 towns and cities, only 21 cities have partial or full sewerage and treatment facilities. Hence, coping with rapid urbanization is a major challenge.

Air and Water Pollution

Majority of industrial plants are using outdated technologies and makeshift facilities devoid of any provision of treating their wastes. A great number of cities and industrial areas have been identified as the worst in terms of air and water pollution. Acts are enforced in the country, but their implementation is not so easy. The reason being the lack of resources, technical expertise, and political and social will. Again, the people are to be made aware of these rules. Their support is indispensable to implement these rules.

Institutions in Environment Conservation

There have been several governmental and non-governmental organizations (NGOs) that have led to a growing interest in environmental protection and conservation of nature and natural resources in the country. The traditional conservation practices that were part of ancient Indian culture have however gradually disappeared. Public awareness is thus a critical need to further environmental protection. Among the large number of institutions that deal with environmental protection and conservation, a few well-known include governmental organizations such as the BSI and ZSI, and NGOs such as BNHS, WWF-I, etc. In the following paragraphs, these are discussed in brief:

Bombay Natural History Society (BNHS), Mumbai

The BNHS began as a small society of six members in 1883. It grew from a group of hunters and people from all walks of life into a major research organization that substantially influenced conservation policy in the country. The influence on wildlife policy building, research, popular publications and people's action has been unique features of the multifaceted society. Undoubtedly, its major contribution has been in the field of wildlife research. It is India's oldest conservation research-based NGO and one that has acted at the forefront of the battle for species and ecosystems. The BNHS publishes a popular magazine called *Hornbill* and also an internationally well-known *Journal on Natural History*. Its other publications include Salim Ali's *Handbook on Birds*, J.C. Daniel's book of *Indian Reptiles*, S.H. Prater's book on *Indian Mammals* and P.V. Bole's book on *Indian Trees*. One of its greatest scientists was Dr Salim Ali whose ornithological work on the birds of the Indian subcontinent is world famous. The BNHS has over the years helped the government to frame wildlife-related laws and has taken up battles such as the 'Save the Silent Valley' campaign.

World Wide Fund for Nature (WWF-I), New Delhi

The WWF-I was initiated in 1969 in Mumbai after which, its headquarters was shifted to Delhi with several branch offices all over India. The early years focused attention on wildlife education and awareness. It runs several programmes and works as a think tank and lobbyist for environment and development issues.

Centre for Science and Environment (CSE), New Delhi

Activities of CSF include organizing campaigns, holding workshops and conferences, and producing environment-related publications. It published a major document on the 'State of India's Environment', which is the first of its kind to be produced as a Citizen's Report on the Environment. The CSE also publishes a popular magazine, *Down to Earth*, which is a science and environment fortnightly. It is involved in the publication of material in the form of books, posters, video films and also conducts workshops and seminars on biodiversity-related issues.

CPR Environmental Education Centre (EEC), Chennai

The CPREEC was set up in 1988. It conducts a variety of programmes to spread environmental awareness and creates an interest in conservation among the general public. It focused attention on NGOs, teachers, women, youth and children to generally promote conservation of nature and natural resources. Its programmes include components on wildlife and biodiversity issues. CPREEC also brings out a large number of publications.

Centre for Environment Education (CEE), Ahmedabad

The CEE, Ahmedabad was created in 1989. It has a wide range of programmes on the environment and produces a variety of educational material. CEE's Training in Environment Education (TEE) programme has trained many environment educators.

Bharati Vidyapeeth Institute of Environment Education and Research (BVIEER), Pune

This is part of the Bharati Vidyapeeth Deemed University. It runs the PhD, the Master's and Bachelor's programmes in Environmental Sciences. It also offers an innovative Diploma in Environmental Education for in-service teachers. Besides a large outreach programme for teachers that has covered over 135 schools, it also conducts fortnightly Environment Education Programmes. Biodiversity conservation is the major focus of its research initiatives. It has developed low-cost Interpretation Centres for Natural and Architectural Sites that are highly locale-specific as well as a large amount of innovative environment educational material for a variety of target groups. Its unique feature is that it conducts environmental education from primary school level to the post-graduate level. The BVIEER has produced several EE aids too. It has developed a teacher's handbook linked to school curriculum, a textbook for UGC for its undergraduate course on environment. It has also developed a CD ROM on India's biodiversity.

Uttarakhand Seva Nidhi (UKSN), Almora

The organization is a nodal agency which supports NGOs in need of funds for their environment-related activities. Its major programme is to organize training for school teachers to use its locale-specific Environment Education Workbook Programme. The main targets are linked with sustainable resource use at the village level through training school children. Its environment education programme covers about 500 schools.

Kalpavriksh, Pune

This NGO, initially Delhi-based, is now working at Pune and is also active in several other parts of India. Kalpavriksh works on a variety of fronts: education and awareness, investigation and research, direct action and lobbying, and litigation with regard to environment and development issues. Its activities include organizing talk and audio-visual shows in schools and colleges, nature walks and outstation camps, ensuring students' participation in ongoing campaigns including street demonstrations, creating consumer awareness regarding organic food, handling green alerts, and meetings with the city's administrators. It is involved in the preparation of site-specific environmental manuals for school teachers. Kalpavriksh was responsible for developing India's National Biodiversity Strategy and Action Plan in 2003.

Salim Ali Centre for Ornithology and Natural History (SACON), Coimbatore

This institution was Dr Salim Ali's dream that became a reality only after his demise. He wished to support a group of committed conservation scientists on a permanent basis. Initially conceived as a wing of the Bombay Natural History Society (BNHS), it later evolved as an independent organization based at Coimbatore in 1990. It has instituted a variety of field programmes that have added to the country's information on our threatened biodiversity.

Wildlife Institute of India (WII), Dehradun

WII was established in 1982 as a major training establishment for research in wildlife management. Its most significant publication has been *Planning a Wildlife Protected Area Network for India* by Rodgers and Panwar published in 1988. The organization has over the years added an enormous amount of information on India's biological wealth. It has trained a large number of Forest Department's officials and staff as Wildlife Managers. Its M.Sc. programme has trained wildlife scientists. It also has an Environment Impact Assessment (EIA) Cell, which trains personnel in eco-development, wildlife biology, habitat management and nature interpretation.

Botanical Survey of India (BSI)

BSI was established in 1890 at the Royal Botanic Gardens, Calcutta. However, it closed down for several years after 1939 and was reopened in 1954. By 1955, the BSI had its headquarters in Calcutta with Circle Offices at Coimbatore, Shillong, Pune and Dehradun. Between 1962 and 1979, offices were established in

Allahabad, Jodhpur, Port Blair, Itanagar and Gangtok. The BSI currently has nine regional centres, which carry out surveys of plant resources in their respective regions.

Zoological Survey of India (ZSI)

ZSI was created in1916. Its mandate was to do a systematic survey of fauna in India. It has over the years collected 'type specimens' on the basis of which animal life has been studied. Its origins were collections based at the Indian Museum at Calcutta, which was established in 1875. Older collections of the Asiatic Society of Bengal, which were made between 1814 and 1875, as well as those of the Indian Museum made between 1875 and 1916, were then transferred to the ZSI. Today, it has over a million specimens. This makes it one of the largest collections in Asia. It has done an enormous amount of work on taxonomy and ecology. It currently operates from 16 regional centres.

Environmental Thinkers and Activists

There are several internationally known environmental thinkers and activists. Among those who have made landmarks, are included Charles Darwin, Ralph Emerson, Henry Thoreau, John Muir, Aldo Leopald, Rachel Carson and E.O. Wilson. Each of these thinkers looked at the environment from a completely different perspective. Charles Darwin wrote the *Origin of Species,* which brought to light the close relationship between habitats and species. It also brought about a new thinking of man's relationship with other species that was based on evolution. Alfred Wallace came to the same conclusions in his work. Ralph Emerson spoke of the dangers of commerce to our environment way back in the 1840s. Henry Thoreau in the 1860s wrote that the wilderness should be preserved after he lived in the wild for a year. He felt that most people did not care for nature and would sell it off for a small sum of money. John Muir is remembered as having saved the great ancient sequoia trees in California'a forests. In the 1890s he formed the Sierra Club, which is a major conservation NGO in the USA. Aldo Leopald was a forest official in the US who in the 1920s designed the early policies on wilderness conservation and wildlife management. In the 1960s, Rachel Carson published several articles that caused immediate worldwide concern on the effects of pesticides on nature and mankind. She wrote a well-known book, called *Silent Spring*, which eventually led to a change in Government policy and public awareness. E.O. Wilson is an entomologist who envisioned that biological diversity was a key to human survival on earth. He wrote *Diversity of Life* in 1993, which was awarded a prize for the best book published on environmental issues. His writings brought home to the world the risks to mankind due to man-made disturbances in natural ecosystems that are leading to the rapid extinction of species at the global level. There have been a number of individuals who have been instrumental in shaping the environmental history in our country. Some of the well-known names in the last century include environmentalists, scientists, administrators, legal experts, educationists and journalists. Salim Ali's name is synonymous with ornithology in India and with the Bombay Natural History Society (BNHS). He also wrote several great books including the famous *Book of Indian Birds*. His autobiography, *Fall of a Sparrow* is a must-read book every nature enthusiast. He was the country's leading conservation scientist and influenced environmental policies in India for over 50 years.

Indira Gandhi as the Prime Minister of India also played a highly significant role in the preservation of country's wildlife. The Wildlife Protection Act was formulated during her period and the Indian Board for Wildlife was extremely active as she personally chaired all its meetings. India gained a name for itself by being a major player in CITES and other international environmental treaties and accords. S.P. Godrej was one of India's greatest supporters of wildlife conservation and nature awareness programmes. Between 1975 and 1999, Godrej received 10 awards for his conservation activities. He was awarded the Padma Bhushan in 1999. His relationship with people in power, combined with his deep commitment for conservation, led to his playing a major advocacy role for wildlife in India. M.S. Swaminathan is one of India's foremost agricultural scientists who has also been concerned with various aspects of biodiversity conservation. He has founded the M.S. Swaminathan Research Foundation in Chennai, which does work on the conservation of biological diversity.

Madhav Gadgil is a well-known ecologist in India. His interests range from broad ecological issues such as developing Community Biodiversity Registers and conserving sacred groves to studies on the behaviour of mammals, birds and insects. He has written several articles, published papers in journals and is the author of six books. M.C. Mehta is undoubtedly India's most famous environmental lawyer. Since 1984, he has filed several public interest litigations (PILs) for supporting the cause of environmental conservation in India. His most famous and long-drawn battles supported by the Supreme Court include protecting the Taj Mahal, cleaning up the Ganges river, banning intensive shrimp farming on the coast, initiating government to implement environmental education in schools and colleges, and a variety of other conservation issues. Anil Agarwal was a journalist who wrote the first report on the 'State of India's Environment' in 1982. He founded the Centre for Science and Environment, which is an active NGO that supports various environmental issues. Medha Patkar is known as one of India's champions who have supported the cause of downtrodden tribal people whose environment is being affected by the dams on the Narmada river. Sunderlal Bahugna's *Chipko Andolan* has become an internationally well-known example of a highly successful conservation action programme with the efforts of local people for guarding their forest resources. His fight to prevent the construction of the Tehri dam in a fragile earthquake-prone setting is a battle that he continues to wage. The Garhwal Hills will always remember his dedication to the cause for which he has walked over 20 thousand kilometres. All above them, the father of the nation Mahatma Gandhi was the forerunner of movements towards conserving the nature. In his writings, he always talked about the nature and its conservation. His famous saying, "the nature can meet the need of the people but not greed of the people" is ever being quoted by the intellectuals and the common people.

CHAPTER 2

Natural Resources

Natural resources refer to all the living and non-living endowments of the earth such as land, water, air, forests, etc. They are undoubtedly the backbone of economic development. On the other hand, the high level of development is due to the availability of natural resources. The areas where availability of natural resources is high, the rate of development is also high. The major classes of natural resources are: land, forests, water, minerals and renewable and non-renewable energy resources. Some natural resources are renewable by natural or artificial processes while others are non-renewable. Currently, the non-renewable resources, such as minerals and fuels, are finite and the rates of their consumption have accelerated in recent decades. Some major renewable resource systems – marine fisheries and some groundwater systems – are being destroyed and that environmental capacities are being seriously exceeded. World petroleum reserves are huge and more are being discovered each year, but these are not located in the major consuming countries. The same is true of bauxite, iron ore, chromium, tin and natural gas. It means that the areas where resources' reserves are tremendous, the consumption rates are considerably low and vice-versa. Therefore, the pattern of development and environmental conditions is variable in all countries. However, the consumption rate of resources is increasing day by day and consequently most of the resources are on the verge of their extinction. This is the case with biodiversity resources also. Many of the floral and faunal resources of the world have been vanished fully and they are registered in red book. It is estimated that when use of a resource grows at 5 per cent per year, it will double in 14 years. Similarly, if the use of known reserves is 100 times more than the current annual use, such reserves will be exhausted in 36 years, and even if a huge discovery doubled reserves to 200 times than the current use, the reserves would last for only 48 years. Under such circumstances, the non-renewable resources will be vanished within a short span of time from the earth.

Definition of Natural Resources

The dictionary meaning of natural resources is a naturally occurring material, i.e., coal or wood that can be exploited by people. According to the United States Geological Survey (USGS), resource is a concentration of naturally occurring solid, liquid, or gaseous material in or on the earth's crust in such form and amount that economic extraction of a commodity from the concentration is currently or potentially feasible. Natural resources are classified into two broad categories: identified and unidentified. The other classification of natural resources is renewable and non-renewable.

Renewable Resources

A resource that can be renewed as quickly as it is used up and can, in theory, last indefinitely is called renewable resource. In the present span of time where population is growing at an alarming rate and natural resources are being exploited largely, the role of renewable resources is noteworthy. The discussion on the renewable resources is as follows:

Biodiversity Resources: Plant and Animal Community

Plant community, predominantly of trees or other woody vegetation, occupying an extensive area of land, is known as forest. In its natural state, a forest remains in a relatively fixed, self-regulated condition over a long period of time. Climate, soil and the topography of the region determine the characteristic trees of a forest. In local environments, dominant species of trees are characteristically associated with certain shrubs and herbs. The type of vegetation on the forest floor is influenced by the larger and taller plants, but because low vegetation affects the organic composition of the soil, the influence is reciprocal. Disturbances such as a forest fire or timber harvesting may result in a shift to another forest type. Left undisturbed, ecological succession will eventually result in a climax forest community. Human intervention is practised to maintain some desirable forest types.

Figure 2.1 Classification of Natural Resources

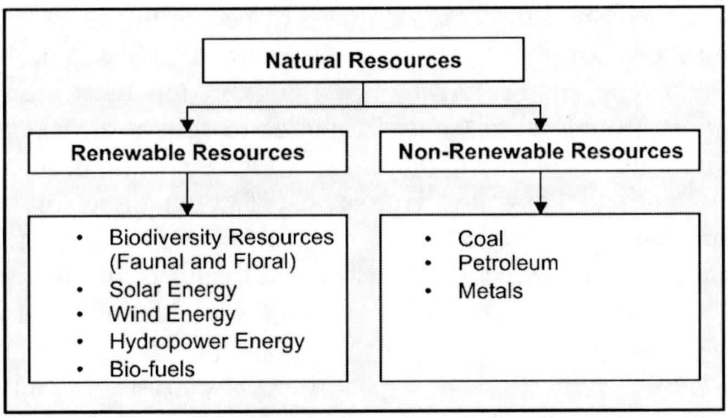

Forests: Classification and Distribution

Forests may be classified into the following eight general types on the basis of leaf characteristics and climate (Encarta, Encyclopedia, 2005):

Tropical (Equatorial) Rainforests

Tropical rainforests are characteristic of Central Africa and the Amazon watershed. Tropical wet forests – the famous rainforests – grow in a great equatorial belt, where conditions for life are as nearly uniform as is possible on land. Temperature, precipitation and day length hardly vary from season to season. The combination of warm temperatures, high rainfall and days of constant length makes for an environment in which plant growth and reproduction is essentially independent of the time of year, so that leaves, flowers, and fruits are always available to feed the animal population. Tree species are highly diverse but usually have smooth, straight trunks and large, simple leaves. The crowns of the rain forest trees form a continuous canopy, above which pop the tops of the occasional emergent giants. Within the forest, the competition for light is intense, and many plants get closer to the sun by scrambling up the larger trees or growing as epiphytes on their branches and trunks.

The wet forests of the Amazon basin are often regarded as the quintessence of rainforests, and indeed they are extraordinarily diverse in both plant and animal species, most of which science has yet to describe. The bloc of Amazon rainforests is the largest on earth, and despite the encroachment of people in many places, it remains comparatively intact. In other parts of the Americas, a ribbon of continuous forest once ran from the Pacific coast of Ecuador through all of Central America to southern Mexico, and another forest, the *mata atlantica,* stretched from northeastern Brazil south to Paraguay. Human use of both these forests has fragmented and degraded them, but the Atlantic forest of Brazil is so critically threatened that it is now the subject of an intensive conservation campaign. Aside from these continental forests, remnant patches of Caribbean rain forest still clothe parts of the Lesser Antilles. In Africa, a western zone of rain forest extends from Guinea to Ghana, where the savannas of the Dahomey Gap separate it from the larger Congo forest in Nigeria, Cameroon, and the countries of the Congo basin. A substantial portion of the Congo forest remains, but the Guinea bloc and the Nigerian forests have been heavily logged, as have the island forests of eastern Madagascar. The most intact wet forests in Asia and Oceania are on the islands of Borneo and New Guinea. Otherwise, tropical rain forest runs in an interrupted swath from the Western Ghats of India to Sri Lanka and Bangladesh, and through Southeast Asia and the Malay Archipelago to northeastern Australia and many of the islands of the tropical Pacific, including Hawaii.

Rainforests are typically hot and steamy. The average annual temperature remains 25°C. Temperature near the equator varies little over the course of a year, so rainforest temperatures are about the same year round – the average minimum monthly temperature in a rain forest is a balmy 18°C. Temperature in a rain forest depends not only on distance from the equator but also on altitude. As elevation increases, night-time temperatures go down significantly. This daily temperature variation affects forest ecology, and true rainforests typically are not found above 1,000 m.

Rainforest ecosystem contains more plant and animal species than virtually any other habitat in the world. Although their range has contracted and expanded with climate changes over the last several million years. In general, rainforests are some of the oldest ecosystems on earth. Rain forest regions are incredibly rich in natural resources and commodities, particularly timber, minerals and petroleum that fetch high prices in international markets. Many trees that grow in rainforests are highly prized for the durability and beauty of their wood. Teak, rosewood and mahogany are rain forest hardwoods used to make furniture and cabinetry all over the world. For many farmers and ranchers, rainforests offer wide expanses of grazing land or land that can be converted for agricultural purposes. Rainforests are also home to a number of exotic animals, such as parrots and monkeys, which bring high prices when captured and sold as pets.

Despite their uniqueness and extraordinary value, tropical rainforests are being destroyed and badly degraded at an unsustainable rate. Some scientists estimate that in the early 1990s tropical forests were being destroyed at a rate of approximately 28 hectares a minute, or about 14 million hectares each year. This figure marked a decrease since the 1980s, when approximately 16 million hectares were destroyed each year, largely due to a reported decline of deforestation in the Amazon river basin in the early 1990s. However, satellite images indicate that rates may have rebounded in the late 1990s as burning in the Amazon increased again. Over the past three decades alone, about 5 million sq. km. or 20 per cent of the world's tropical forests have been cleared. During this time, deforestation in tropical Asia reached almost 30 per cent. High rates of deforestation are inevitably followed by alarming rates of plant and animal extinction because many rain forest species cannot survive outside their pristine rain forest habitat. Some scientists estimate that dozens of rain forest species are becoming extinct every day.

Box 2.1 Conserving Rainforest

Since the 1970s, an increasing number of national and international organizations have been established to promote rain forest conservation. In recent years, two principal approaches have been followed to conserve tropical rainforests – strict protection and sustainable development. The first, outright protection through the development of national parks and preserves has been an essential element in biodiversity conservation. In this method of conservation, entire tracts of the rain forest are set aside, and uses are carefully regulated. Protected areas are especially important for preserving the most biologically distinctive rainforests, that is, those forests with exceptionally high species diversity with many species endemic to that forest. At the end of the 20th century, the number of national parks in tropical countries increased significantly, and today approximately 5 per cent of all forests are classified under some protected status. International organizations, such as the World Bank and the World Wildlife Fund (WWF), have launched a movement in all countries in the developing world to set aside 10 per cent of their forests in protected areas.

Tropical Dry Forests and Savannas

Savannas are tropical grassland with a scattering of shrubs and small and large trees. Savannas may result from soil conditions, from periodic fires caused by lightning or set by humans, or from climatic influences. Climatically determined savannas, as found in western and southwestern Africa, develop in regions with marked wet and dry seasons, where rainfall ranges between 100 and 400 mm. a year. These savannas vary from open-canopied forests with a grassy understory to true savannas in which grasses are dominant. When the rainfall is 100 to 200 mm., generally only grasses can survive. When rainfall reaches 300 mm., the soil holds enough water to sustain shrubs through the dry season as well. Tropical savanna forests are found in regions such as the campos of Brazil, where forest and grassland meet. Savannas, which occur widely in Africa and South America, are dominated by grasses and sedges, with open stands of widely spaced trees that are frequently thorny.

Some savannas are created by fire or by grazing and browsing mammals. Much more typical are the deciduous woodlands and savannas that occupy those regions of the tropics in which a rainy season alternates with a dry one. Just as trees in temperate zone forests cast their leaves during the drought imposed by the cold winter, deciduous trees in the tropics shed their leaves during the protracted dry season. As a consequence, the same tropical forest that appears so lush and green during the rains assumes a sere and lifeless aspect during the months of drought. If the amount and pattern of rainfall were the only determinants of the vegetation, a continuous cover of deciduous forests would cloak the terrain that lies between the rainforests and the deserts of the tropics. In fact, the closed forests are often replaced with open woodlands, or by savannas in which the trees are spaced even more widely over a carpet of grass. In some places, the trees are confined to strips along the course of dry streams, and the country resembles pure grassland.

The savannas and woodlands of Africa are renowned for the large and diverse herds of hoofed mammals they support. The comparatively moist Guinea savannas and woodlands, with their tall grasses and many trees, surround the rainforests on the north, east and south. With increasing aridity, these savannas give way to the Sudan savannas and woodlands of the Sahel, the Horn of Africa, and the Kalahari, parts of which qualify as semi-deserts. In Asia, deciduous monsoon forests cover much of India and Indo-China. Northern Australia contains similar woodlands and savannas. Deciduous dry forests line much of the Pacific coast of Mexico and Central America. Similar forests cover most of Cuba, the Yucatan peninsula and western Ecuador. The llanos of Venezuela and Colombia are a region of extensive savanna grasslands, parts of which flood during the rainy season. Enormous tracts of central Brazil, the campos cerrados, are covered by tropical woodlands. Farther to the south are the vast, marshy pantanal and the savannas of the Gran Chaco, which Brazil shares with Argentina, Bolivia and Paraguay. In regions of higher rainfall, such as eastern Africa, savanna vegetation is maintained by periodic fires. Consuming dry grass at the end of the rainy season, the fires burn back the forest vegetation, check the invasion of trees and shrubs, and stimulate new grass growth. Woody vegetation then increases, changing savanna to woodland. Such woody growth can be reduced, in turn, by large browsers such as elephants.

Deciduous Monsoon Forests

Deciduous monsoon forests are characteristic of Bengal and Myanmar and common throughout Southeast Asia and India. They are also found along the Pacific coastal regions of Mexico and Central America. The climate is characterized by heavy daily rainfall, seasonally relieved by dry periods during which the trees shed their leaves.

Temperate Deciduous Forests

Forests composed of deciduous species, which periodically shed their leaves, grow in well-watered parts of the temperate zone in which a long, warm growing season alternates with a cold winter. Precipitation, as either rain or snow, tends to fall throughout the year. Deciduous trees, shrubs and woody vines, whether in the tropics or in the middle latitudes, drop their leaves during the time of drought. Temperate deciduous forests are almost entirely confined to the Northern Hemisphere, where they occur in three major tracts. In Europe, the zone of deciduous and mixed forests stretches from the British Isles and France through central and eastern Europe as far as the Ural mountains. In East Asia, these forests clothe the Russian Far East, Korea and Japan. Their counterparts in North America occupy most of the area from the Great Lakes to the Atlantic Ocean and south to the Gulf of Mexico. Although widely separated, these deciduous forests are much alike, not only in appearance but also in the groups of plants that compose them: birches, hornbeams, alders, beeches, oaks, chestnuts, lindens, elms, walnuts, maples and ashes. A long history of human occupation of these wooded regions, especially in Eurasia, has reduced many forests to tiny remnants. In parts of Western Europe, forest destruction have resulted in the development of extensive areas of treeless heaths.

Temperate and Sub-tropical Evergreen Forests

Temperate evergreen forests are found in the sub-tropical regions of North America and the Caribbean islands that have a warm maritime climate. The type is best developed along the Gulf coast and in the Florida Everglades. The characteristic trees are oak, magnolia, palms and bromeliads. In the middle latitudes, the prevailing winds carry moisture air masses over the west coasts of the continents, which receive substantial rainfall as a consequence. In much the same manner, the sub-tropical east coasts of the continents are kept wet by the trade winds. In these regions, the combination of ample precipitation and comparatively mild temperatures produces dense forests in which evergreen species – broadleaf and coniferous – preponderate. In the temperate evergreen forests of the west coasts, rainfall is concentrated in the cool winter months. The mild summers are comparatively dry. These forests line the coast of North America from central California north to southeastern Alaska. The largest trees are all conifers, but broadleaf trees and shrubs are important components. The famous redwoods that dominate the forests of the foggy California coast are replaced farther north with spruces, hemlocks, and other giant conifers. Despite its similar climate, the much-interrupted coast of northwestern Europe lacks this type of evergreen forest, which was eliminated during the last period of glaciations. In the Southern Hemisphere, temperate rainforests of southern beeches, podocarps and other species are well developed in central Chile and in Tasmania and the South Island of New Zealand. The most extensive region of sub-tropical evergreen forests is in East Asia, where rain falls throughout the year and summers are hot. These forests occupy a large transitional zone between the deciduous forests of northern China and Japan and the tropical evergreen forests of Southeast Asia. Far to the south of the equator, significant sub-tropical evergreen forests grow in southeastern and southwestern Australia, New Zealand, and the coast and highlands of southern Brazil.

Mediterranean Woodlands

Temperate rainforests, with broad-leaved evergreen trees, are common in Mediterranean coasts. Rainfall may be low, but the ocean-cooled air is moisture-laden, and fogs are frequent. In the United States, the temperate West Coast rainforests are dominated by hemlock, cedar, spruce, fir and redwood. A climate of mild, wet winters and hot, dry summers produces characteristic woodlands composed of evergreen shrubs and small trees, many with small, hard leaves – sclerophyll – that resist drying by closing the pores through which water vapour can escape. Fire is a regular feature of these woodlands, whose scrubby character is maintained by frequent burns. Many species in these woodlands are fire-resistant, such as the thick-barked Mediterranean cork oak. Mediterranean woodlands grow in five widely scattered regions. The best known, and the ones for which this class of eco-regions is named, are the woodlands that grow all around the Mediterranean Sea. Almost all of this area's original vegetation of evergreen oaks and pines has been altered by people over the course of the last several millennia. Woodlands similar to the maquis occupy the region of the Cape of Good Hope in South Africa. Called the fynbos, this eco-region contains an extraordinary number of species, many found nowhere else. In southwestern Australia, eucalyptus trees and protea shrubs dominate the jarrah and wandoo forests that occupy the zone of Mediterranean climate. In Chile, south of the desert, is a sclerophyllous scrub community called the matorral, and in central and southern California, the coastal hills are covered with the manzanitas, ceanothus, scrub oaks, and other plants that make up the chaparral. A type of shrub-land community that is dominated by small-leaved evergreen vegetation is called chaparral. Such habitats are characteristic of the Mediterranean type of climate with warm, wet winters and long, dry summers. The name is applied to the shrub lands of California and Baja California that are dominated by scrub oak and by the dense shrubs chamiso and manzanita. Chaparral is fire dependent. In other areas with a Mediterranean climate, equivalent plant communities are given such local names as the *tomillares* of Spain, the *macchia* of the Mediterranean countries and South Africa, the *phrygana* of the Balkans, and the brigalow shrub of South Australia.

Boreal Evergreen (Taiga) Forests

Northern coniferous forests form a worldwide belt in sub-arctic and alpine regions of the Northern Hemisphere. Gnarled scrub trees dominate at the northern tree line and on mountaintops. Spruce and fir trees are characteristic of the more northerly forests; pine, larch and hemlock dominate farther south. These forests usually occupy formerly glaciated regions and occur in association with lakes, bogs and rivers. A great band of evergreen conifer forests extends around the Northern Hemisphere in the sub-Arctic latitudes. Within this zone, the growing season is brief and the winter is long and cold. These circumstances exclude most broadleaf deciduous trees, which require more time to re-grow their leaves than the sub-arctic climate provides. Compared with the forests of lower latitudes, the boreal forests, often called by their Russian name of 'taiga', are rather uniform in structure and species composition. Ground vegetation within the shady forest is sparse, comprising low shrubs, many of them in the heath family, and a layer of mosses. Another conspicuous feature of the taiga region, from which the ice sheets retreated only about 10,000 years ago, is the countless ponds and bogs that pock the flat and poorly drained terrain.

In Eurasia, the taiga spans the continent from the Russian Far East to Scandinavia. The westernmost forests are very monotonous, but the number of species rises as one moves to the east, and the forests of Hokkaido in northern Japan are comparatively diverse. An exception to the evergreen character of most of the taiga is the deciduous larch forests of eastern Siberia, where the bitterly cold winters of the continental interior make the retention of any leaves, even needles, and a liability.

Tundra Forests

Tundra, Arctic plains, encompassing most of earth's terrain north of the coniferous forest belt, dominated by cotton grass, heath, lichen, moss, sedge and willow. Similar plains, called alpine tundra, occur above the timberline in the high mountains of the world. The Antarctic region has a few areas of tundra as well. Tundra climate is characterized by harsh winters, low average temperatures, little snow or rainfall and a short summer season. The Arctic tundra, in particular, is influenced by permafrost – a layer of permanently frozen sub-soil in the ground. The surface soil, which tends to be rocky, thaws in summer to varying depths. The combination of frozen ground and flat terrain on the tundra impedes the drainage of water. Common to the alpine tundra is bare rock-covered ground, called fell-fields, supporting a growth of lichens. The numerous microhabitats provided by these landforms give variety to the tundra landscape. The number of plant species on the tundra is few and their growth is low, with most of the biomass concentrated in the roots. Typical arctic vegetation comprises cotton grass, sedge and dwarf heath, together with associated mosses and lichens. Arctic wildlife is circumpolar; the same or closely related species are found around the world. The variety of animal life is also limited in the challenging environment. Musk-ox, caribou and reindeer are the dominant large grazers, feeding on grass, sedge, lichen and willow. Arctic hare, or snowshoe rabbit, and lemming feed on grass and sedge. Predators include the wolf, arctic fox and snowy owl. Polar bears and sometimes brown bears are seen. Many birds nest in the tundra shrubbery in summer, migrating to milder climates before the winter season sets in. Invertebrate life is scarce, but insects such as black flies and mosquitoes are abundant. Alpine animal life includes the mountain goat, big-horned sheep, pika, marmot, and the ptarmigan – a grouse like bird. Flies are scarce but butterflies, beetles and grasshoppers are abundant.

Box 2.2 The Tundra Ecosystem

The tundra ecosystem is extremely sensitive to disturbance with little ability to restore itself. Here, wildlife is also vulnerable to habitat destruction, to over hunting, and to extinction through loss of any of the animal or plant species that make up the fragile, highly interdependent tundra community of life. Recent studies have demonstrated that global warming is impacting areas of tundra, particularly the Arctic tundra. Impacts include the melting of permafrost and the expansion of the timberline, which in turn decreases the areas of tundra. Permafrost melting contributes to this feedback loop by releasing methane, a greenhouse gas, from the soil into the atmosphere. The expanded timberline results in more sunlight being absorbed by trees, which warms the earth's surface. Trees have a low albedo as compared with the tundra. The tundra tends to reflect sunlight back into space, especially when it is

Cont'd...

...Cont'd

snow-covered and has a high albedo, thereby cooling earth's surface. An expanded timberline and decreased areas of tundra therefore lead to more heat being absorbed at the surface. Shrinking areas of tundra also have a negative impact on the animal life that depends on tundra vegetation. Some studies have indicated that areas of tundra are disappearing more rapidly than expected due to global warming. Recent studies have demonstrated that global warming is impacting areas of tundra, particularly the Arctic tundra. Impacts include the melting of permafrost and the expansion of the timberline, which in turn decreases the areas of tundra. Both impacts contribute further to global warming, creating a so-called positive feedback loop that makes it more difficult to halt the warming trend. Permafrost melting contributes to this feedback loop by releasing methane, a greenhouse gas, from the soil into the atmosphere. The expanded timberline results in more sunlight being absorbed by trees, which warms the earth's surface. Trees have a low albedo compared with the tundra. The tundra tends to reflect sunlight back into space, especially when it is snow-covered and has a high albedo, thereby cooling earth's surface. An expanded timberline and decreased areas of tundra therefore lead to more heat being absorbed at the surface.

Solar Energy

Solar energy is a major source of renewable energy that can solve the energy problem if used scientifically. Energy radiated from the sun in the form of heat and light is called solar energy. On the other hand, radiation produced by nuclear fusion reactions deep in the sun's core is called solar energy. The sun provides almost all the heat and light earth receives and therefore sustains every living being. Solar energy travels to earth through space in discrete packets of energy called photons. On the side of earth facing the sun, a square kilometre at the outer edge of our atmosphere receives 1,400 megawatts of solar power every minute, which is about the capacity of the largest electric-generating plant in Nevada. Only half of that amount, however, reaches earth's surface. The atmosphere and clouds absorb or scatter the other half of the incoming sunlight. The amount of light that reaches any particular point on the ground depends on the time of day, the day of the year, the amount of cloud cover, and the latitude at that point. The solar intensity varies with the time of day, peaking at solar noon and declining to a minimum at sunset. The total radiation power (1.4 kilowatts per square metre, called the solar constant) varies only slightly – about 0.2 per cent every 30 years. Any substantial change would alter or end life on earth.

Wind Energy

A renewable energy – wind energy – contained in the force of the winds blowing across the earth's surface. When harnessed, wind energy can be converted into mechanical energy for performing works such as pumping water, grinding grain and milling lumber. By connecting a spinning rotor to an electric generator, modern wind turbines convert wind energy, which turns the rotor into electrical energy. Wind is created when air that has been warmed over sun-heated land rises, leaving a vacuum in the space it once occupied. Cooler surrounding air then rushes in to fill the vacuum. This movement of rushing air is what we know as wind. Egyptians may have been the first to capture wind energy when they sailed boats up the Nile River beginning around the 4th century BC. For centuries afterward, wind-powered sailing vessels plied the world's seas and oceans, serving as the principal form of commercial transport. Wind energy has been harnessed on land since the first windmill was developed by the ancient Persians in the 7th century AD. Windmills have since been used to mill grain, pump water, saw timber and provide other forms of mechanical energy. As wind is a clean and renewable source of energy, modern wind turbines had been installed in 26 countries by 2007, including such nations as Germany, Denmark, India, China and the United States, to supplement more traditional sources of electric power, such as burning coal. Design improvements, such as more efficient rotor blades combined with an increase in the numbers of wind turbines installed, have helped increase the world's wind energy generating capacity by nearly 150 per cent since 1990. In 2006 the United States became the world's third largest producer of energy from wind power, generating more than 11.5 megawatts of electricity.

Box 2.3 Evolution of the Windmill

The term ‘windmill’ derives from using these machines to grind, or mill, grain. The oldest known windmills were crude, simple devices used in the 7th century by the Persians. Europeans made extensive use of the windmill beginning in the 12th century, providing mechanical energy for pumping water, sawing lumber, and grinding grain. In the late 20th century, windmills were developed to convert wind energy into electric power. Early Persian windmills were crude devices consisting of a simple tower supporting an array of paddles made from bundled reeds. The first windmills to appear in Europe were built during the 12th century in northwestern France and southern England. Use of the windmill subsequently spread into northern Belgium, Germany, and north to Denmark during thc late 12th and 13th centuries. European windmill performance increased greatly over the next 500 years. During the zenith of the European windmill, some 1,500 megawatts of power were being produced, a level not reached again until 1988. The American farm windmill, ideally suited for pumping water from deep underground, became an integral part of agricultural communities across the American West. A series of improvements were made during the 100-year-reign of the American water-pumping windmill. The oil crisis of the 1970s spurred efforts into developing wind energy as an alternative source of electrical power. Many countries launched programmes to develop modern wind turbines. While many of these programmes failed, Denmark was successful in developing modern wind turbines. The modern wind turbine is the result of design and material advances made during the 1980s and 1990s, which have enabled wind turbines to become increasingly efficient. Awareness of global warming also spurred interest in developing wind energy during the late 20th and early 21st centuries. In many locations wind energy offered an attractive alternative to fossil fuels, which release carbon dioxide, a greenhouse gas, into the atmosphere.

Hydroelectricity

Hydroelectricity is generated by converting the pressure of falling or running water to electricity by means of turbine fastened to a generator. Waterpower is derived from the fall of water from a higher to a lower level, and extracted by means of waterwheels or hydraulic turbines. Waterpower is a natural resource, available wherever a sufficient volume of steady water flow exists. The development of waterpower today requires extensive construction, including storage lakes, dams, bypass canals, and the installation of large turbines and electric generating equipment. As the development of hydroelectric power requires a large capital investment, it is often uneconomical for a region where coal or oil is cheap, even though the cost of fuel for a steam-powered generating plant is higher than the cost of running a hydroelectric plant. However, increasing environmental concerns are focussing attention on renewable energy sources such as water.

Box 2.4 History of Hydroelectricity Power

The use of waterpower dates from ancient Greece and Rome, where waterwheels were used for the milling of corn. The availability of cheap slave and animal labour, however, restricted the widespread application of waterpower until about the 12th century. During the middle ages, large wooden waterwheels were developed with a maximum power output of about 50 horsepower. Modern waterpower owes its development to the British civil engineer John Smeaton, who first built large waterwheels using cast-iron construction. Waterpower played an important part in the development of early American towns. Starting with the grist mill and sawmill of colonial times, waterpower gave impetus to the growth of the textile, leather, and machine-shop industries in the early 1800s. Most of the cities in the Midwest developed along the so-called fall line, where the upland mountain region meets the coastal plain. Combining the advantages of available waterpower with location at the upstream end of river navigation, the earliest waterworks that led to the growth of cities in the northeast were located at what are now Pawtucket, Rhode Island (1790), Paterson, New Jersey (1791), and Fall River, Massachusetts (1813). Large storage-dam construction, however, was not feasible, and low water flows during summer and fall, coupled with icing during the winter, led to the replacement of nearly all waterwheels by steam when coal became readily available. In the Himalayan region traditional water mills have been using for centuries are widespread in the entire region.

Development of Hydroelectricity Power

The rebirth of waterpower had to await the development of the electric generator, further improvement of the hydraulic turbine, and the growing demand for electricity by the turn of the 20th century. Commercial power companies began to instal a large number of small hydroelectric plants in the mountain regions near the major population centres. By 1920 hydroelectric plants accounted for 40 per cent of the electric power produced in the United States. Motivated by the search for the multiple use of water resources, including navigation, flood control and irrigation, in addition to power production, the Tennessee Valley Authority (TVA) started government participation in large-scale waterpower development in 1933. Most major installations depend on a large water-storage reservoir upstream of the dam where water flow can be controlled and a nearly constant water level can be assured.

Small-scale hydroelectric plants with capacities between 1 kilowatt and 1 megawatt are being developed in some countries. In many of China's districts, for example, such dams are the main source of electric power. Other developing nations are also showing interest in such projects, which can make good use of available labour. In the United States, interest in small-scale plants increased following passage of the Public Utilities Regulatory Policies Act in 1978, because it states that large utilities must buy power fed into their lines by these small-scale plants. In the Indian Himalayan region, small and large hydropower plants have been installed in numerous places.

Global Hydroelectric Power Generation

Worldwide, hydropower represented 17 per cent of the total energy generated in 2003 – the most recent year for which data are available. In many countries, hydroelectric power is the dominant source of electric power. In 2003, Norway derived 99 per cent of its power from hydroelectric plants. The same year, hydroelectric power provided 100 per cent of the electricity used in the Democratic Republic of the Congo and 84 per cent of the electricity used in Brazil. The Itaipu hydroelectric plant on the Parana river between Brazil and Paraguay, officially dedicated in 1982, has the greatest capacity in the world (12,600 megawatts when placed in full operation). The Grand Coulee dam, the largest hydroelectric power plant in North America, provides about 6,480 megawatts. Canada, the largest producer of hydroelectric power in the world, generated 332.5 billion kilowatt-hours (kwh) in 2003. This figure constituted 59 per cent of the nation's electric power. Hydroelectric power generation in the United States increased from about 16 billion kwh in 1920 to 275.8 billion kwh in 2003. Although the United States runs a close second to Canada in the total amount of hydroelectric power produced, only 7 per cent of the electric power used in the United States was generated by hydroelectric power plants in 2003. In India, Tehri dam is the biggest dam in Asia which generates 3,000 megawatts electricity. There are mushrooming of hydropower plants.

Environmental Impact of Hydropower Project

Hydroelectric power is regarded by some as a relatively clean source of energy because it emits fewer greenhouse gases than thermal power plants. Greenhouse gases contribute to global warming and climate change, and for many environmentalists, the build-up of greenhouse gases in the atmosphere is the most important environmental issue. However, the dams used to generate hydroelectricity can have other adverse environmental consequences. Dams alter the water temperatures and micro-habitats downstream. Water released from behind dams usually comes from close to the bottom of the reservoirs, where little sunlight penetrates. This frigid water significantly lowers the temperatures of sun-warmed shallows downstream, rendering them unfit for certain kinds of fish and other wildlife. Natural rivers surge and meander, creating small pools and sandbars that provide a place for young fish, insects, and other river-dwelling organisms to flourish. But dams alter the river flow, eliminating these micro-habitats and, in some cases, their inhabitants. Finally, dams also prevent nutrient-laden silt from flowing downstream and into river valleys. Water in a fast-moving river carries tiny particles of soil and organic material. When the water reaches a pool or a flat section of a river course, it slows down. As it slows, the organic matter it carries drops to the river bottom or accumulates along the banks. Following heavy rains or snowmelt, rivers spill over their banks and deposit organic matter on their floodplains, creating rich, fertile soil. Some of the organic matter makes it all the way to river mouths, where it settles into the rich mud of estuaries, ecosystems that nourish up to one-half of the living matter in the world's oceans. Large dams artificially slow water to a near standstill, causing the organic matter to settle to the bottom of the reservoir. In such cases, downstream regions are deprived of nutrient-laden silt.

Biofuel

Biofuel is a renewable fuel that is derived from biological matters. Fuel made from organic material produced by living things is called biofuel. Biofuel includes any solid, liquid, or gaseous fuel produced either directly from plants or indirectly from organic industrial, commercial, domestic, or agricultural wastes. In principle, burning biofuels adds less carbon to the environment than burning fossil fuels because the carbon atoms released by burning biofuel already existed as part of the modern carbon cycle. Burning fossil fuels, on the other hand, always adds extra carbon because the carbon they contain comes from a buried source that was not part of the modern carbon cycle. Carbon dioxide is the main greenhouse gas thought to contribute to global warming. Biofuels are seen as one way to reduce the amount of carbon dioxide gas added to the atmosphere. Plants used for biofuels take up the same amount of carbon dioxide as they grow as was released by burning biofuels made from an earlier crop. Biofuel is produced and used within the same country is way to reduce dependence on foreign sources of oil and other fuels, providing energy security and an economic boost for agriculture and industry. Biofuels are also a type of renewable energy resource, unlike fossil fuels, which cannot be grown or created.

Types and Uses of Biofuels

Biofuels are mainly made from plant and plant-derived substances known as biomass. The energy in biomass originally comes from the sun through photosynthesis – a process that harnesses sunlight to turn carbon dioxide and water into sugars. Plants then use the sugars to create carbohydrates and cellulose that can later be turned into fuel. There are three main methods for the development of biofuels: (1) the burning of dry organic wastes, (2) energy forestry and (3) the fermentation of wet wastes (such as animal dung) in the absence of oxygen to produce biogas. Firewood remains a primary energy source in many countries. In industrialized countries, biomass is usually processed into liquid fuels that can power combustion engines. The main liquid biofuels are ethanol and bio-diesel. To produce ethanol, carbohydrates in biomass are turned into sugars, which are then fermented. Other plant sources include wheat, rice, sorghum, sunflower, potatoes and sugar beets. Biomass can also be converted into bio-oil. Its plant material such as sawdust or sugarcane residue is heated to a temperature of around 400°C to 500°C (752°F to 932°F) in an oxygen-free environment for less than two seconds. The dark brown liquid that results can be burned as a fuel in boilers that generate electricity, though other uses may be feasible in the future.

Box 2.5 Potential Negative Effects of Biofuel Production

Politicians, scientists, and economists have promoted biofuels as renewable energy sources that add less carbon to the environment and provide energy independence. However, some experts have raised concerns about possible negative consequences of growing crops to provide biofuels. Food prices may rise if a major percentage of grain crops are grown for energy and if areas of food crops are converted to energy crops. The price increases could affect impoverished populations in developing countries, in particular. Rainforests and other tropical areas could be cleared to cultivate crops such as oil palm and soybeans, at the same time displacing populations in regions such as Southeast Asia and the Amazon. India's main environmental concern is its growing population, which is expected to increase to 1.8 billion by the year 2050. In order to feed such a large population, more groundwater will be needed to irrigate crops, increasing the risk of poor soil quality due to salinization. More artificial fertilizer will likely be applied to crop fields, posing threats to drinking water. The demand for meat has increased with greater levels of prosperity, resulting in overgrazing and increasing wasteland. The demand for fuelwood has grown with rural populations, leading to the loss of trees and forests. To decrease reliance on fuelwood, the government has promoted the use of biogas for cooking fuel.

The most widely used liquid biofuel in industrial countries is ethanol. In the United States, in 2005, 3.9 billion gallons of ethanol were produced, mainly from corn. Most of that ethanol was blended with gasoline, accounting for 2.8 per cent of total yearly gasoline sales. The European Union (EU) produced 718 million gallons of ethanol in 2005, mostly distilled from sugar beets and wheat grain. Brazil is the world's largest producer of ethanol. About 15 per cent of its liquid fuel is ethanol derived from sugarcane. Gasoline sold in Brazil is usually mixed with 25 per cent ethanol. Use of biodiesel is also increasing, often blended with diesel derived from petroleum. In 2005, the EU produced 858 million gallons of biodiesel, primarily from rapeseed oil. The United States produced about 250 million gallons of biodiesel in 2006, nearly all of it from soybean oil.

Non-Renewable Resources

Non-renewable resources are those resources which cannot be renewed once used. Most of the world energy need is supported by non-renewable resources. Alongwith overuse of these resources, they are becoming non-existential. Coal and petroleum will be vanished within 40 years according to an estimation. Similarly, the other non-renewable resources will also be exhausted if they will continue to be used unabated. Non-renewable resources are discussed as follows (Encarta, Encyclopedia, 2005):

Coal

Coal, a combustible organic rock, composed primarily of carbon, hydrogen and oxygen, when burns produces energy. It is also used to manufacture steel. It is an important source of chemicals used to make medicine, fertilizers, pesticides, and other products. Coal is rich in hydrocarbons. All life forms contain hydrocarbons, and in general, material that contains hydrocarbons is called organic material. Coal originally formed from ancient plants that died, decomposed, and were buried under layers of sediment during the Carboniferous period, about 360 million to 290 million years ago. As more and more layers of sediment formed over this decomposed plant material, the overburden exerted, increasing heat and weight on the organic matter. Over millions of years, these physical conditions caused coal to form from the carbon, hydrogen, oxygen, nitrogen, sulfur, and inorganic mineral compounds in the plant matter. Plant matter changes into coal in stages. In each successive stage, higher pressure and heat from the accumulating overburden increase the carbon content of the plant matter and drive out more of its moisture content. Scientists classify coal according to its fixed carbon content. Higher grades of coal have higher fixed carbon content.

Box 2.6 Coal Formation

Coal is a sedimentary rock formed from plants that flourished millions of years ago when tropical swamps covered large areas of the world. Its formation began during the Carboniferous period (first coal age), which spanned 360 million to 290 million years ago. Coal formation continued throughout the Permian, Triassic, Jurassic, Cretaceous and Tertiary periods (second coal age), which spanned 290 million to 1.6 million years ago. Coals formed during the first coal age are older, so they are generally located deeper in earth's crust. The greater heat and pressures at these depths produce higher-grade coals such as anthracite and bituminous coals. Conversely, coals formed during the second coal age under less intense heat and pressure is generally located at shallower depths. Consequently, these coals tend to be lower-grade subbituminous and lignite coals.

Modern Uses of Coal

Eighty-six per cent of the coal used is burnt by electric power plants to produce electricity. When burnt, coal generates energy in the form of heat. As stated above, it is also used in the steel industry. Other industries use different coal gases given off during the coke-forming process to make fertilizers, solvents, medicine, pesticides, and other products. Fuel companies convert coal into easily transportable gas or liquid fuels. Coal-based vapour fuels are produced through the process of gasification. Gasification may be accomplished either at the site of the coalmine or in processing plants. In processing plants, the coal is heated in the presence of steam and oxygen to produce synthesis gas, a mixture of carbon monoxide, hydrogen, and methane used directly as fuel or refined into cleaner-burning gas. On-site gasification is accomplished by controlled, incomplete burning of an underground coal bed while adding air and steam. To do this, workers ignite the coal bed, pump air and steam underground into the burning coal, and then pump the resulting gases from the ground. Once the gases are withdrawn, they may be burned to produce heat or generate electricity. Or they may be used in synthetic gases to produce chemicals or to help create liquid fuels. Liquefaction processes convert coal into a liquid fuel that has a composition similar to that of crude petroleum. Coal can be liquefied either by direct or indirect processes. However, because coal is a hydrogen-deficient hydrocarbon, any process used to convert coal into liquid or other alternative fuels must add hydrogen. Four general methods are used for liquefaction: (1) pyrolysis and hydrocarbonization, in which coal is heated in the absence of air or in a stream of hydrogen; (2) solvent extraction, in which coal hydrocarbons are selectively dissolved and hydrogen is added to produce the desired liquids; (3) catalytic liquefaction, in which hydrogenation takes place in the presence of a catalyst; and (4) indirect liquefaction, in which carbon monoxide and hydrogen are combined in the presence of a catalyst.

Coal Deposit and Reserves

Although coal deposits exist in nearly every region of the world, commercially significant coal resources occur only in Europe, Asia, Australia and North America. Commercially significant coal deposits occur in sedimentary rock basins typically sandwiched as layers called beds or seams between layers of sandstone and shale. When experts make estimates of the world's coal supply, they distinguish between coal reserves and resources. Reserves are coal deposits that can be mined profitably with existing technology – that is, with current equipment and methods. Resources are an estimate of the world's total coal deposits, regardless of whether the deposits are commercially accessible. Exploration geologists have found and mapped the world's most extensive coal beds. At the beginning of 2001, global coal reserves were estimated at 984.2

billion metric tons (1 metric ton equals 1,016 kg.). These reserves occurred in the following regions by order of importance: the Asia Pacific, including Australia, 29.7 per cent; North America, 26.1 per cent; Russia and the countries of the former Union of Soviet Socialist Republics (USSR), 23.4 per cent; Europe, excluding the former USSR, 12.4 per cent; Africa and the Middle East, 6.2 per cent; and South and Central America, 2.2 per cent. In India coal reserves are basically found in four states – Bihar (including Jharkhand), West Bengal, Madhya Pradesh (including Chhattisgarh) and Orissa. Coal deposits in the United Kingdom, which led the world in coal production until the 20th century, extend throughout parts of England, Wales and southern Scotland. Coalfields in Western Europe underlie the Saar and Ruhr valleys in Germany, the Alsace region of France and areas of Belgium. Coalfields in Central Europe extend throughout parts of Poland, the Czech Republic and Hungary. The most extensive and valuable coalfield in Eastern Europe is the Donets basin, between the Dnieper and Don rivers (in parts of Russia and Ukraine). Large coal deposits in Russia are being mined in the Kuznetsk basin in southern Siberia. Coalfields underlying northwestern China are among the largest in the world. Mining of these fields began in the 20th century. United States coal reserves are located in six major regions, three of which produce the majority of domestically mined coal. The most productive region in the United States is the Appalachian basin, covering parts of Pennsylvania, West Virginia, Kentucky, Tennessee, Ohio and Alabama. Large quantities of coal have also been produced by both the Illinois basin – extending through Illinois, Indiana and Kentucky – and the Western Interior Region – extending through Missouri, Kansas and Oklahoma. Other commercially important US coal regions include the Powder River basin, underlying parts of Montana and Wyoming; the Green River basin in Wyoming; the Uinta basin, covering areas of Utah and Colorado; and the San Juan basin, underlying parts of Utah, New Mexico and Colorado. In 2006, estimates of total US coal reserves were approximately 17 billion metric tons. At the beginning of the 21st century production amounted to about 980 million metric tons each year.

Petroleum

Petroleum, or crude oil, naturally occurring oily, bituminous liquid, is composed of various organic chemicals. It is found in large quantities below the surface of earth and is used as a fuel and as a raw material in the chemical industry. Modern industrial societies

Box 2.7 Historical Development and Exploration

These surface deposits of crude oil have been known to humans for thousands of years. By the time the Renaissance began in the 14th century, some surface deposits were being distilled to obtain lubricants and medicinal products, but the real exploitation of crude oil did not begin until the 19th century. The Industrial Revolution had by then brought about a search for new fuels, and the social changes it effected had produced a need for good, cheap oil for lamps; people wished to be able to work and read after dark. The search for a better lamp fuel led to a great demand for "rock oil" that is, crude oil – and various scientists in the mid-19th century were developing processes to make commercial use of it. Thus, the quest for greater supplies of crude oil began. The first such wells were dug in Germany from 1857 to 1859, but the event that gained world fame was the drilling of an oil well near Oil Creek, Pennsylvania, by Colonel Edwin L. Drake in 1859. With the invention of the automobile and the energy needs brought on by World War I (1914–1918), the petroleum industry became one of the foundations of industrial society. Petroleum geologists and geophysicists have many tools at their disposal to assist in identifying potential areas for drilling. Thus, surface mapping of outcrops of sedimentary beds makes possible the interpretation of subsurface features, which can then be supplemented with information obtained by drilling into the crust and retrieving cores or samples of the rock layers encountered. In fact, most of the oil provinces in the world have initially been identified by the presence of surface seeps, and most of the actual reservoirs have been discovered by so-called wildcatters who relied perhaps as much on intuition as on science. An oil field, once found, may comprise more than one reservoir – that is, more than one single, continuous, bounded accumulation of oil. Most of the oil that has been discovered and exploited in the world has been found in a relatively few large reservoirs.

use it primarily to achieve a degree of mobility – on land, at sea, and in the air – that was barely imaginable less than 100 years ago. In addition, petroleum and its derivatives are used in the manufacturing of medicines and fertilizers, foodstuffs, plastics, building materials, paints and cloth, and to generate electricity. In fact, modern industrial civilization depends on petroleum and its products; the physical structure and way of life of the suburban communities that surround the great cities are the result of an ample and inexpensive supply of petroleum. In addition, the goals of developing countries – to exploit their natural resources and to supply foodstuffs for the burgeoning populations – are based on the assumption of petroleum availability. In recent years, however, the worldwide availability of petroleum has steadily declined and its relative cost has increased. Many experts forecast that petroleum will no longer be a common commercial material by the mid-21st century. Petroleum is the world's major source of energy and is a key factor in the continued development of world economies. There is a growing awareness worldwide of the need for a consistent set of resources and reserves definitions for use by governments and industry in the classification of petroleum reserves. The experience of recent decades shows that the increment of available resources exceeds the consumed part of those resources, giving a positive balance favourable to the well-being of future generations. In other words, currently, there is no scarcity problem regarding non-renewable resources.

Production Volumes and Reserves

Crude oil is perhaps the most useful and versatile raw material that has become available for exploitation. By 2003, the United States was using 7 billion barrels of petroleum per year, and worldwide consumption of petroleum was 29.3 billion barrels per year. The world's technically recoverable reserves of conventional crude oil amounted to about 1.3 trillion barrels in 2004, of which some 216 billion barrels were in North America. Of the known oil reserves that can be profitably extracted at current prices, about 57 per cent are in the Middle East. However, in the 21st century, oil companies have started focusing more on exploiting unconventional crude oil reserves. Improved techniques for the extraction of unconventional crude oil, especially of bitumen in the tar sand deposits of western Canada, combined with higher conventional crude oil prices, made exploitation of unconventional reserves more cost-efficient. The reserves of bitumen in the province of Alberta were believed sufficient to yield 1.7 trillion barrels of synthetic crude oil. Only 10 per cent of this amount, however, was thought to be technically recoverable. Nevertheless, by 2007, Canada had become the principal foreign supplier of oil to the United States. In India crude oil reservoirs are located in the Western Coast – Gujarat, Maharashtra and Bombay High. It is also found in the northeastern regions, particularly in Assam valley.

Environmental Effects of Using Petroleum

Using petroleum products for development activities is a major cause of global warming. Petroleum combustion releases carbon dioxide, a greenhouse gas, into the atmosphere, and rising levels of greenhouse gases are driving to climate changes. These conditions could cause numerous environmental problems, including disrupted weather patterns and polar ice cap melting. Disrupted weather patterns could lead to extensive drought and desertification. Polar ice cap melting could cause flooding and profound changes in ocean circulation. Many environmental organizations are urging governments and individuals to reduce greenhouse gas emissions by conserving energy with fuel-efficient technologies and by developing alternative, renewable energy sources such as wind and solar power. In the United States, most environmental groups have urged the government to ratify the Kyoto Protocol – a global treaty that sets a specific timetable for reducing greenhouse gas emissions. Drilling oil wells also creates environmental problems because the petroleum pumped up from deep reservoir rocks is often accompanied by large volumes of salt water. The extraction of unconventional crude oil sources, such as bitumen, poses even greater environmental problems due to the large number of pollutants produced and the use of fossil fuels involved in the extraction. Petroleum usually must also be transported long distances by tankers or pipelines to reach a refinery. Transportation of petroleum occasionally leads to accidental spills. Oil spills, especially in large volumes, can be detrimental to wildlife and habitat.

Metal Resources

Metals are non-renewable resources and are classified as precious and non-precious metals. The consumption of metal resources is increasing in all over the world and many of the metals are on the verge of disappearing. It is said that if these trends of exploitation and consumption are allowed to continue unchecked, the reserves of metals will be completely vanished in the coming future.

Table 2.1 Evolution of the Reserves and Identified Resources during the Period 1950–2000

	1950	*1974*	*2000*	*Growth in 50 years (2000/ 1950)*
Mineral Commodity	*Reserves and Identified Resources*	*Reserves and Identified Resources*	*Reserves and Identified Resources*	
Bauxite	1.4 × 10 9	1.6 × 10 10	3.5 × 10 10	25.0×
Copper	1.0 × 108	3.9 × 108	6.5 × 10 8	6.5×
Gold	3.1 × 10 4	4.0 × 10 4	7.7 × 10 4	2.5×
Iron	1.9 × 10 10	8.8 × 10 10	3.1 × 10 11	16.3×
Lead	4.0 × 107	1.5 × 10 8	1.3 × 10 8	3.25×
Nickel	1.4 × 107	4.4 × 10 7	1.5 × 10 8	10.7×
Phosphate rock	2.6 × 109	1.3 × 10 10	3.7 × 10 10	14.2×
Silver	1.6 × 10 5	1.9 × 10 5	4.2 × 10 5	2.6×
Tin	6.0 × 106	1.0 × 107	12.0 × 10 6	2.0×
Zinc	7.0 × 107	1.2 × 10 8	4.3 × 10 8	6.1×
Coal	6.0 × 1010	6.5 × 1011	9.8 × 10 11	16.3
Oil (bbl)	8.0 × 10 10	7.2 × 10 11	1.05 × 10 12	13.1×
Natural Gas (m3)	4.7 × 1012	2.2 × 10 15	1.5 × 10 14	31.9×

Source: USGS, BP Statistical Review of World Energy. The Petroleum Handbook, Royal Dutch/Shell Group of Companies, Fifth edition, 1966, IEA (several), UN-Statistical Yearbook

Figure 2.2 Major Forms of Metal Resources

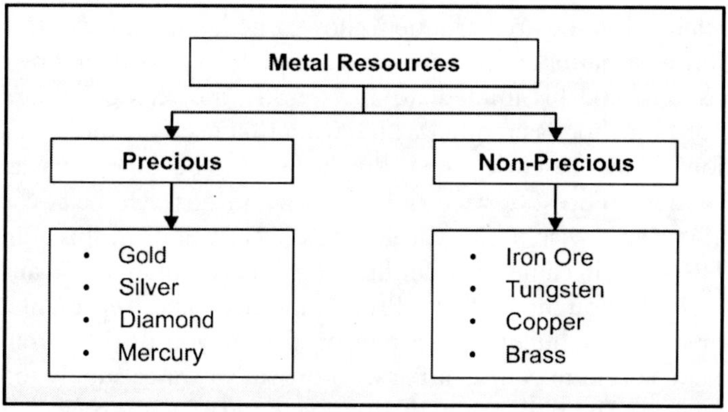

Conservation and Sustainability

It could have been predicted that the modern conservation movement would have its beginnings not in the settled lands of the Old World but in those areas of the New World where, within the memory of a single generation, there had been extreme changes in the landscape and in the abundance of wildlife. The reaction to the destruction of natural resources in those areas precipitated the formation and growth of the conservation movement. As early as 1832, George Catlin, a US artist and author, first proposed the idea of national parks encompassing major areas. In the same decade botanist William Bartram and ornithologist John James Audubon were arousing an interest in wildlife and its conservation. A little later, the famous writers Ralph Waldo Emerson and Henry David Thoreau presented strong arguments concerning the importance of the continued survival of wild nature to the psychological well-being of mankind. Thoreau became one of the first literary advocates of wilderness conservation. The first textbook on conservation, *Man and Nature*, by George Perkins Marsh, appeared in the 1860s. In the same period the author and naturalist John Muir settled in California and became a leading advocate of wilderness preservation. In 1872 the US Congress proclaimed the Yellowstone region of Wyoming as a national park and also established for the first time a national government's role in the protection and administration of such areas. In 1891 the first of the US forest reserves, forerunners of the system of national forests, was proclaimed in the area around Yellowstone National Park. Conservation as a national movement owes much to President Theodore Roosevelt, and his immediate advisers. Roosevelt's chief forester, Gifford Pinchot, is credited with having first used the term "conservation" in its present context. Pinchot was to become the leader of the nation's Forest Service and, along with Roosevelt, he advocated a utilitarian, "wise use" approach to conservation. In this, Pinchot and Roosevelt came into conflict with representatives of another school of thought, called "Preservationist" and represented by Muir, over the building of a dam in the canyon of the Tuolumne river in Yosemite National Park, to provide a water supply for San Francisco. The philosophy of Roosevelt and Pinchot prevailed and the dam was built, but Muir's ideas lived on with the Sierra Club, of which he was a founder.

The recent history of conservation has been marked by a great expansion of governments roles in protecting the environment and by a growth of public interest in and support for this process. National park systems, dedicated to the preservation of wild nature and to the provision of outdoor recreation space, have grown rapidly, and national forest systems, dedicated to the multiple uses of wild-land resources, have also become firmly established. In the United States the conservation of wildlife became a cause of national interest and led to the establishment of a far-ranging system of wildlife refuges and the gradual restoration of most wild animal species to levels approaching, in some cases exceeding, their primitive abundance. On private lands, however, and on government or public-domain lands not specifically reserved as national forests, parks, or refuges, deterioration continued, reaching a peak in the 1930s, when it became widely recognized that those rangelands in the public domain had been disastrously overgrazed and that many privately-owned farmlands had been depleted or exhausted. Firm control over the management of lands in the public domain and federal intervention to establish soil conservation on privately owned lands were accepted as appropriate activities for the national government.

Conservation ideas spread widely, being most readily accepted by those countries that had experienced sudden environmental changes. By the 1920s national parks were to be found in all continents. In 1924 the Soviet Union established the first of its now extensive system of natural reserves. Conservation-oriented management of forestlands, which grew more from its origins in Europe than from practices in the United States, also became more widely accepted throughout the world. The scientific basis for the management of wild grazing lands for the sustained production of forage for livestock was established in US national forests in 1913 and soon spread to other countries. Aldo Leopold in the United States, in 1933, wrote a textbook on game management, in which the conservation and management of wild animal life for such recreational purposes as sport hunting and fishing and for direct commodity values on a sustained basis received particular emphasis. Leopold's work drew heavily on earlier studies of animal ecology by Charles Sutherland Elton in England; in fact, the establishment in Europe of wildlife reserves and protective laws as well as the managing of lands to produce sustained crops of wildlife long preceded even Elton's work. Subsequently, the management of wild animals in extensive wilderness areas made major strides in Africa, which possesses unusual wildlife resources, and in the Soviet Union, which retains large areas of wild land.

Regarding minerals, resource conservation and environmental changes are closely related. Both concepts deal with scarcity or even the complete exhaustion rather in the economic sense than in the physical sense. Simultaneously, the environment suffers an impact, as mineral bodies themselves are part of the environment. In certain examples, exhaustion and environmental change can be understood as necessary and acceptable; in other cases, one or both, must be considered as waste and degradation. The issues to be analysed are when, where and for whom each case applies, and these questions only can be resolved in terms of trade-offs. To assess the trade-offs, one must compare what was gained and what was lost because of production and usage of minerals. Such comparisons cannot be made only in physical units, but should incorporate economic units and, in many cases social units too. In this discussion, some issues are relevant:

- What quantity of a non-renewable resource shall be produced and consumed in a given period of time?
- How can the best use of these non-renewable resources produced in that period of time be achieved?
- How can the renewable resources associated to non-renewable in nature be better protected?

Resources Availability: Scarcity Views

Concerns over the availability of mineral resources needed to maintain the standard of living dated back to the 18th century, when Thomas Malthus observed the land and resources expanding at an arithmetic rate while population grew at a geometric rate. However, this concern has been sporadic, tending to arise when the markets for resources are tight and prices high, and to decrease when these markets are weak and prices low. Two schools of thought have been exploring various aspects of resource availability over the last decades. First, the *empirical school* focuses on measuring trends in resources availability over time. This approach requires an appropriate measure of scarcity, for example, physical measures, such as parts per million of a metal in the earth's crust or the resource base. In addition to physical measures, there are economic measures, of which three are commonly used: real price, cost of production for the marginal producer, and the value of mineral in the ground prior to the development of the deposit in which it is found. The economists sometimes refer to this as the pure rent associated with a deposit, which under certain conditions can be similar to the cost of discovery. Both physical and economic measures are accepted, but they may be misleading, if not used appropriately.

The best-known study within the empirical school is *Scarcity and Growth* (Barnett and Morse, 1963). These authors assess the trends from 1870 to 1957 for different commodities, suggesting that the resources in general are becoming less than rather scarcer, despite their extensive exploitation over time. The principal factor appears to have been the introduction of new technologies, whose cost-reducing effects over the last century apparently more than offset the cost-increasing effects of resource depletion. For these reasons, this approach emphasizes the opportunity cost of finding and processing exhaustible resources, rejecting the fixed stock paradigm. The second approach, called *theoretical school*, determines the optimal use of a fixed stock resource over time and is represented by the contribution of Hotelling (1931), and a voluminous body of subsequent papers. Hotelling's major contribution was to demonstrate that a firm exploiting a fixed resource stock would behave differently from other firms operating in other economic sectors. A firm exploiting a non-renewable resource needs to recognize that current production incurs in two types of costs. The first covers production costs and is similar in nature to the marginal cost considered by all firms. The second is the opportunity cost associated with depleting the fixed resource stock and thus reducing the quantity available for future use. This user cost, as it is commonly called, is the present value of the reduction in future profits foregone by producing one more unit of output during this period.

Another important aspect developed by Hotelling has an important implication in resource scarcity theory that over time the value of a resource left in the ground will rise at the firm's rate of time discount, which reflects the rate of interest that the firm can earn on other investment opportunities. In another form, this finding seems intuitively, because if the value of the resource is not growing at such rate, the firm has an incentive to mine the mineral resource as quickly as possible and place the profits in other higher yielding investments. This finding indicates that resource scarcity will increase over time, because the value of resource in the ground is one of the economic measures of scarcity. This is very controversial, because the literature provides little support for this conclusion of the Hotelling theory, and in fact, finds considerable evidence to the contrary. Some authors criticize the Hotelling approach because we cannot employ the principle of a fixed stock by defining it as the economic portion, estimated under uncertainty about quality and cost. For example, if ultimate production of oil and gas is determined by future costs and prices, the estimated resource cannot be the starting point for estimating costs prices. What actually exist are flows from undiscovered resources into a reserve inventory. This rationale is based upon that there are good reasons to expect reserve replenishments to show diminishing returns over time. As we can see, the estimations of reserves and identified resources for the last fifty years show crescent growth values for the majority of commodities (Encyclopedia of Life Support Systems).

Natural resources – renewable and non-renewable – are the backbone of the modern economy. Due to high level of development all over the world, large-scale exploitation of resources has occurred during the last 50 years or so and consequently, many of them are on the verge of vanishing. Keeping the unavailability of non-renewable resources, the countries are looking towards renewable resources for their potential utilization.

CHAPTER 3

Conservation of Natural Resources

Natural resource conservation can be defined as sustainable use and protection of natural resources including plants, animals, mineral deposits, soils, clean water, clean air, and fossil fuels such as coal, petroleum and natural gas. As discussed in the preceding chapter, natural resources are grouped into two categories – renewable and non-renewable. A renewable resource is one that may be replaced over time by natural processes, such as fish population or natural vegetation, or is inexhaustible, such as solar energy, wind energy, biofuel and hydropower. The objective of renewable resource conservation is to ensure that such resources are not consumed faster than they are replaced. Non-renewable resources are limited and cannot be replaced or can be replaced only over extremely long periods of time. Non-renewable resources include fossil fuels – coal and petroleum and mineral deposits – iron ore and gold ore. Conservation activities for non-renewable resources focus on maintaining an adequate supply of these resources in the future.

Natural resources are conserved for not only their biological, economic and recreational values, but also their natural beauty and importance to local cultures. For example, tropical rainforests are protected for their important role in both global ecology and the economic livelihood of the local culture; a coral reef may be protected for its recreational value for scuba divers; and a scenic river may be protected for its natural beauty. Conservation conflicts arise when natural resource shortages develop in the face of steadily increasing demands from a growing human population. Controversy frequently surrounds how a resource should be used, or allocated, and for whom. For example, a river may supply water for agricultural irrigation, habitat for fish, and water-generated electricity for a factory. Farmers, fishermen, and industry leaders vie for unrestricted access to this river, but such freedom could destroy the resource, and conservation methods are necessary to protect the river for future use. It also worsens when a natural resource crosses political boundaries. For example, the headwaters, or source, of a major river may be located in a different country than the country through which it flows. There is no guarantee that the river source will be protected to accommodate resource needs downstream. In addition, the way in which one natural resource is managed has a direct effect upon other natural resources. Cutting down a forest near a river, for instance, increases erosion, the wearing away of topsoil, and can lead to flooding. Eroded soil and silt cloud the river and adversely affect many organisms such as fish and important aquatic plants that require clean, clear freshwater for survival (Encarta, Encyclopedia, 2005).

Historical Perspectives

Conservation of natural resources has long been practising in India since time immemorial as it is based on religion and culture and the people are virtually habitual for conserving all natural resources. Here, the people worshipped trees and animals. Trees of Peepal, Bargad (banyan) and many others have been worshipping from time immemorial. Similarly, cow is worshipped as we call it 'Mother Cow'. Actually, in Hinduism, the earth is itself called as 'Mother'. We have fairs, festivals, culture and customs in the name of trees and animals. In the distant past, particularly at the time of Mahabharata, the practices of conserving trees and animals in the form of worshipping them, were flourishing. Lord Krishna, while preaching to Arjuna in the battle of Kurukshetra, says that '*Pedon main main Peepal hun or gayon main Kamdainu*' that means I am Peepal among trees and Kamdhenu among cows. In the 1970s and

1980s, courageous grassroots organizations such as the Chipko movement in India (a coalition of villagers, mostly women) and the Brazilian rubber tappers (workers who extract chicle, the tree sap used to make rubber) fought for preservation of the forests that provided their livelihood.

Until the advent and spread of Christianity and Islam in the 4th and 5th centuries, there were many religions based on animism – the belief that all objects have a spiritual being. This belief led to careful protection of natural resources out of fear or respect for these spiritual beings. Moreover, early agricultural lifestyles, dependent on nature to provide good crops and growing conditions, also encouraged sound land use practices. Ancient Phoenicians, Greeks and Romans developed irrigation, crop rotation and terraced hillsides as early methods of water, nutrient and soil conservation.

In Europe, the relationship between humanity and nature became strained with the beginning of the Industrial Revolution in the 19th and early 20th centuries. Industrialization stifled traditional agricultural lifestyles and encouraged urbanization and innovation in science and technology to control nature and extract resources. The Industrial Revolution led to environmental damage on a grand scale as European technology spread around the globe. Coal burning and iron smelting produced waste that contaminated air and water, the concentrated population in urban areas produced huge amounts of unconfined raw sewage that contaminated drinking water, and vast forests and plains were cleared for agriculture. The European conservation movement began to grow as the effects of industrialization worsened in the mid-20th century. Clean air legislation was enacted in the United Kingdom in 1956 in reaction to London's industrial smog, which killed more than 2,000 people in early December 1952. Political parties with environmental or conservation agendas sprang up in New Zealand, Australia and Europe by the 1970s, and became known as 'Green Parties' in the 1980s.

The modern conservation movement of the United States began in the mid-19th century when resource depletion and pollution were first becoming serious problems. As conservation ideas gained support, a wave of conservation activity swept the country. The world's first national park, Yellowstone National Park, was established in Wyoming in 1872 to protect an area of incredible natural beauty. The conservation movement rose into the spotlight again in the 1960s as publications, such as *Silent Spring* (1962) by American biologist Rachel Carson, raised public concerns about the health and environmental hazards of pesticides and other toxic chemicals used by industry. Several catastrophic events in 1969, including the toxic waste fires on the Cuyahoga river in Cleveland, Ohio and a coastal oil spill in Santa Barbara, California, focused media attention on the need for environmental conservation. The estimated 20 million people across the United States who attended the first national Earth Day, a day for recognizing environmental concerns, on April 22, 1970, demonstrated massive public support for conservation issues. Conservation legislation passed in the 1970s included the Endangered Species Act, the Marine Mammal Protection Act, the Clean Air Act, the Clean Water Act, and the Toxic Substance Control Act.

The 1980s experienced a slowdown of the conservation momentum of the 1970s. Resource conservation concerns remained in the public mind, however, due to continued scientific discoveries concerning global warming, acid rain and depletion of the ozone layer. Ecological disasters such as the Bhopal gas tragedy in 1984, nuclear reactor explosion near the Ukrainian town of Chernobyl in 1986 and the tanker Exxon Valdez oil spill in Prince William Sound, Alaska in 1989, served as catastrophic reminders of the effects of human carelessness.

In 1972, the United Nations Environment Programme was formed to encourage international cooperation in conservation and development strategies. Collaboration on environmental conservation issues included the 1987 Montreal Protocol to protect the ozone layer, the 1992 United Nations Conference on Environment and Development (UNCED) in Rio de Janeiro, Brazil, and the 1994 United Nations Conference on Population and Development in Cairo, Egypt. The participation of the United States in this international movement was unwilling, while Canadian and European support and participation were strong. The 1992 UNCED conference, commonly referred to as the Earth Summit, was the largest international meeting ever held with 178 nations participating. Its proceedings noted the economic and environmental gulf between the northern and southern hemispheres and emphasized a sustainable growth, utilitarian approach to conservation. In the same year an appeal, entitled *World Scientists' Warning to Humanity,* was released. This paper was signed by 1,700 of the world's leading scientists, including 104 Nobel laureate scientists, 19 national academies of science, and the director general of the United Nations Educational, Scientific and Cultural Organization (UNESCO). It stated that at the current rate of consumption, the earth's resources may soon be reduced to the point at which the living world would be "unable to sustain life in the manner that we know". In several publications, ecologists and economists agree that despite the immediate benefits of economic growth, infinite growth in material and energy consumption is not compatible with the finite resources of the earth and will undermine the well-being of both economic and ecological systems. For these reasons, natural resource conservation has become one of the most important challenges to face the human race.

How Can Natural Resources Be Conserved?

The challenge of conservation is to understand the complex connections among natural resources and balance resource use with protection to ensure an adequate supply for future generations. In order to accomplish this goal, a variety of conservation methods are used. These include reducing consumption of resources; protecting them from contamination or pollution; reusing or recycling resources when possible; and fully protecting, or preserving resources. Consumption of natural resources rises dramatically every year as the human population increases and standards of living rise. From 1950 to 2000 the world population reached more than double, with nearly 80 per cent living in developing, or poorer, nations. The large, developed nations, however, are responsible for the greatest consumption of natural resources because of their high standards of living. For instance, on average, an American consumes as much energy as 27 Filipinos or 370 Ethiopians consume. Conservation education and the thoughtful use of natural resources is essential in the developed countries to reduce their consumption. For example, reduction in the demand for tropical hardwoods such as teak and mahogany in the United States and Japan has slowed down the rate of tropical forest destruction considerably.

To protect natural resources from pollution, individuals, industries and governments have many obligations. These include prohibiting or limiting the use of pesticides and other toxic chemicals, limiting wastewater and airborne pollutants, preventing the production of radioactive materials, and regulating drilling and transportation of petroleum products. Failure to do so has resulted in contaminated air, soil, rivers, plants and animals. For example, if governments restricted all oil tankers to be fitted with double-layered hulls, the damages to fisheries and wildlife from many oil spills of the 20th century, such as the 1967 Torrey Canyon oil spill in the English Channel, could have been reduced. In many cases it is possible to reuse or recycle resources to reduce waste and resource consumption and conserve the energy needed to produce consumer products. For example, paper, glass, *freon* (a refrigerant gas), aluminum, metal scrap and motor oil can all be recycled. A preventative measure called *precycling*, a general term for designing more durable, recyclable products such as reusable packaging, encourages reuse. Many provinces in the United States have established mandatory recycling laws in an attempt to reduce waste and consumption.

Some resources are so unique or valuable that they are protected from activities that would destroy or degrade them. For example, national parks and wilderness areas are protected from logging or mining because such activities would reduce the economic, recreational and aesthetic values of the resource. Forests and wetlands may be protected from development because they enhance air and water quality and provide habitat for a wide variety of plants and animals. Unfortunately, these areas are often threatened with development because it is difficult to measure the economic benefits of cleaner air, cleaner water, and the many other environmental benefits of these ecosystems.

Types of Conservation

There is a variety of basic conservation methods used to protect global natural resources. Although, each resource has a unique set of conservation problems and solutions, all resources are interconnected in a complex and little-understood web. Scientists have learnt that damaging one thread of the web may weaken the entire structure. It is important that this connectivity be addressed in the search for solutions to resource shortages. It would be impractical to work towards the conservation of soil, for instance, without considering the needs and effects of nearby water and vegetation resources.

Biodiversity Conservation

Biodiversity, or biological diversity, denotes the number and variety of different organisms and ecosystems in a certain area. Preserving biodiversity is essential for ecosystems to respond flexibly to damage or change. For example, a single-species corn crop may be quickly destroyed by a certain insect or disease, but if several different species of corn are planted in the field, some of them may resist the insect or disease and survive. The same principle applies to natural areas, which adapt to natural environmental changes such as wildfire, drought, or disease because of the biodiversity that has evolved in the area over thousands, or even millions, of years. Some trees, such as lodge pole pine, may even require fire to aid in reproduction. These trees produce cones that are opened by extreme heat. The fire opens the cones and the seeds are then released into the soil.

Figure 3.1 Forms of Resource Conservation

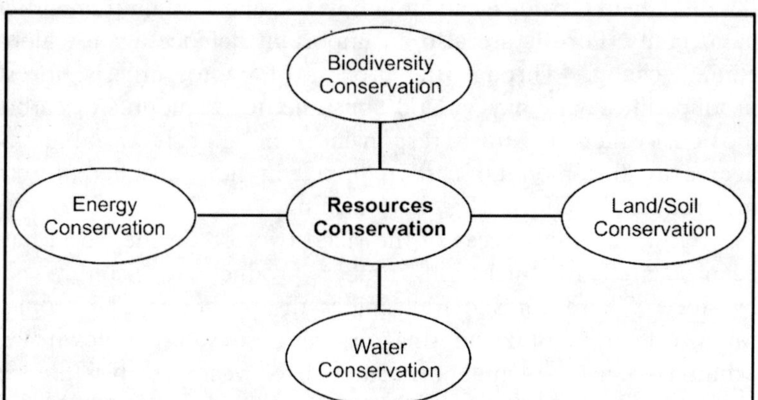

Humans benefit greatly from many medicines, crops, and other products that biodiversity provides. About 40 per cent of our modern pharmaceutical medicines are derived from plants or animals. Unfortunately, human activities have greatly reduced biodiversity around the world. The 20th century encompasses one of the greatest waves of extinction, or elimination of species, to occur on the planet. The greatest threat to biodiversity is loss of habitat as humans develop land for agriculture, grazing livestock, industry and habitation. The most drastic damage has occurred in the tropical rainforests, which cover less than 7 per cent of the earth's surface but contain well over half of the planet's biodiversity. Only 8 per cent of the rainforests in Madagascar, home of the rosy periwinkle – remain intact. Several nations have laws protecting endangered species. An international treaty – the Convention on International Trade in Endangered Species of Wild Fauna and Flora (CITES) – went into effect in 1975 and outlawed trade of endangered animals and animal parts. In India various measures are adopted to conserve endangered species. A new scientific field – conservation biology – studies ways to stop the destruction of biodiversity and restore natural habitats.

Forest Conservation

Forests provide many social, economic and environmental benefits. In addition to timber and paper products, forests provide wildlife habitat and recreational opportunities, prevent soil erosion and flooding, help provide clean air and water, and contain tremendous biodiversity. Forests are also an important defence against global climate change. Through the process of photosynthesis, forests produce life-giving oxygen and consume huge amounts of carbon dioxide, which is most responsible for global warming. By decreasing the amount of carbon dioxide in the atmosphere, forests may reduce the effects of global warming.

However, huge areas of the richest forests in the world have been cleared for fuelwood, timber products, agriculture and livestock. These forests are rapidly disappearing. The tropical rainforests of the Brazilian Amazon river basin were cut down at an estimated rate of 14 million hectares each year, which is an area about the size of the state of Wisconsin (USA) in the 1990s. The countries with the most tropical forests tend to be developing and overpopulated nations in the southern hemisphere. Due to poor economies, people resort to clearing the forest and planting crops in order to survive. While there have been effective efforts to stop deforestation directly through boycotts of multinational corporations responsible for exploitative logging, the most effective conservation policies in these countries have been efforts to relieve poverty and expand access to education and health care.

In 2005, the Food and Agriculture Organization (FAO) of the United Nations issued a major report, titled “Global Forest Resources Assessment 2005,” on the status of the world’s forests. Based on a five-year study, the report found that forested areas throughout the world were continuing to decline at a rate of about 7.3 million hectares per year – an area equivalent in size to Panama or Sierra Leone. However, the rate of decline had slowed in comparison with the period from 1990 to 2000, when the world lost about 8.9 million hectares of forested area per year. Africa and South America continued to have the largest net loss of forests, while forest loss also continued in North and Central America and the Pacific Islands. Only Europe and Asia showed a net gain in forested areas due to forest planting, landscape restoration and expansion of natural forests. China, in particular, reported a large-scale afforestation effort. But, in India, forest cover dramatically declined about 2 per cent after independence. In 2005, the world’s total forest area was just less than 4 billion hectares.

Worldwide, the nations are working for conserving forest resource. In India, Forest Act of 1982 was passed to conserve the forests. Apart from this, several other Acts were framed and implemented such as Wild Life Act, Environmental Protection Act, etc. Large-scale afforestation was carried out to check soil erosion and increase the net forest cover. But, the increase in forest area is negative.

Soil Conservation

Soil, a mixture of mineral, plant and animal materials, is essential for most plant growth and is the basic resource for agricultural production. Soil-forming processes may take thousands of years, and are slowed by natural erosion forces such as wind and rain.

Humans have accelerated these erosion processes by developing the land and clearing away the vegetation that holds water and soil in place. The rapid deforestation taking place in the tropics is especially damaging because the thin layer of soil that remains is fragile and quickly washes away when exposed to the heavy tropical rains. Globally, agriculture accounts for 28 per cent of the nearly 2 billion hectares of soil that have been degraded by human activities; overgrazing is responsible for 34 per cent; and deforestation is responsible for 29 per cent.

In addition to reducing deforestation and overgrazing, soil conservation involves reforming agricultural soil management methods. Some of the most effective methods include strip cropping, alternating strips of crop and uncultivated land to minimize erosion and water runoff; contour farming, planting crops along the contours of sloping lands to minimize erosion and runoff; terracing, which also reduces erosion and runoff on slopes; growing legumes, such as clover or soybeans, to restore essential nitrogen in the soil; and minimizing tillage, or plowing, to reduce erosion.

Mountains of the world are worse affected by soil erosion as about 20 per cent of the world population lives in the mountain area. In the Indian Himalayan region, the state governments are working towards soil conservation.

Water Conservation

Clean freshwater resources are essential for drinking, bathing, cooking, irrigation, industry, and for plant and animal survival. Unfortunately, the global supply of freshwater is distributed unevenly. Chronic water shortages exist in most of Africa and drought is common over much of the globe. The sources of most freshwater supplies – groundwater, reservoirs, and rivers – are under severe and increasing environmental stress because of overuse, water pollution and ecosystem degradation. Over 95 per cent of urban sewage in developing countries is discharged untreated into surface waters such as rivers and harbors. About 65 per cent of the global freshwater supply is used in agriculture and 25 per cent is used in industry. Freshwater conservation therefore requires a reduction in wasteful practices like inefficient irrigation, reforms in agriculture and industry, and strict pollution controls worldwide. In addition, water supplies can be increased through effective management of watersheds (areas that drain into one shared waterway). By restoring natural vegetation to forests or fields, communities can increase the storage and filtering capacity of these watersheds and minimize wasteful flooding and erosion. Restoration and protection of wetlands is crucial to water conservation. Like giant sponges, wetlands stabilize groundwater supplies by holding rainfall and discharging the water slowly, acting as natural flood-control reservoirs.

Energy Conservation

All human cultures require the production and use of energy, i.e., resources with the capacity to produce work or power. Energy is used for transportation, heating, cooling, cooking, lighting and industrial production. The world energy supply depends on many different resources including traditional fuels such as firewood and animal waste, which are significant energy sources in many developing countries. Fossil fuels account for more than 90 per cent of global energy production but are considered problematic resources. They are non-renewable means, they can be depleted, and their use causes air pollution. In particular, coal plants have been one of the worst industrial polluters since the beginning of the Industrial Revolution of the 19th century. Moreover, mining or drilling for fossil fuels has caused extensive environmental damage.

There is a global need to increase energy conservation and the use of renewable energy resources. Renewable alternatives, such as waterpower, solar energy, wind energy and geothermal energy, are efficient and practical but largely underutilized because of the ready availability of inexpensive, non-renewable fossil fuels in industrial countries. While some countries, such as France and Japan, depend heavily on nuclear energy, it is still not a major energy source. Excessive production costs, serious safety concerns, and problems with the handling of the dangerous radioactive wastes have virtually eliminated it as a viable energy source. In addition to using alternative energy resources such as solar and wind power, energy conservation measures include improving energy efficiency. For instance, transportation accounts for most of the oil consumption in the world countries. Encouraging the expansion and use of public transportation systems and car-pooling dramatically increases energy efficiency. In the household, energy can be conserved by turning down thermostats, switching off unnecessary lights, insulating homes, and using less hot water.

Conservation of natural resources is essential and is the need of the hour, particularly in the wake of high consumption and over exploitation of non-renewable resources. The non-renewable resources are limited and if the current trend or exploitation of resources is to be continued, the petroleum products and coal would be vanished upto 2040. Biodiversity has been reduced, forests worldwide depleted, high level of soil erosion in many forms has already taken place, water crisis is everywhere and it has a worse situation, a cause of conflict between the state and countries and energy is being reduced at a large-scale. All these impediments together manifested a way for global crisis of natural resources as already many countries lacking in natural resources reserves. Conservation of natural resources is required to be considered at every level with maximum use of renewable resources.

CHAPTER 4

Watershed Management

'Watershed' is a land area that drains into a particular lake, river and oceans while 'management' refers to protection of land, water and biomass. Thus, watershed management can be defined as the conservation of land, water and biomass of a drain area. Watershed, on the other hand, is defined as a geo-hydrological unit draining to a common point by a system of drains. All lands on earth are part of one watershed or other. Watershed is also considered as a biological, physical, economic and social system. The life-sustaining components – land, water and biomass – are the crucial actors in watershed management. It is the right time to educate the people for permanent solution to attain 'water prosperity' and it lies in the adoption of watershed management practices. Incomplete and improper management of the resources has a cumulative effect leading to degradation of environment, low productivity, low income and lack of sustainability in development. Thus, watershed is accepted to be the 'natural home' for managing the life-sustaining components, such as land, water and biomass, in a sustainable manner.

Watershed Management

Watershed management has emerged as a new paradigm for planning, development and management of land, water and biomass resources with a focus on social and environmental aspects following a participatory approach. Watershed management is more a philosophy of comprehensive integrated approach to natural resources management. It aims at integration of social resources management with natural resources management. The approach is generally preventive, progressive, corrective and curative. Watershed management involves the judicious use of natural resources with active participation of institutions/organizations in harmony with the ecosystem.

The management of a single unit of land with its water drainage system is called watershed management. It is a technique that has several components. This includes soil and water management and developing vegetative cover. The natural drainage pattern of a watershed unit if managed appropriately can bring about local prosperity by a year-round abundance of water that improves the quality of human life in the area. As it provides clean water throughout the year, this improves health in the community. Watershed management enhances the growth of agricultural crops and even makes it possible to grow more crop than needed in a year in dry areas. Watershed management begins by taking control over a degraded site through local participation. People must appreciate the need to improve the availability of water both in quantity and quality for their own area. Once this is adequately demonstrated, the community begins to understand to work together in the activities that lead to good watershed management. The first technical step is to take appropriate soil conservation measures. This is done by constructing a series of long trenches and mounds along contours of the hill to hold the rainwater and allow it to percolate into the ground. This ensures that underground stores of water are fully recharged. This is enhanced by allowing grasses and shrubs to grow and by planting trees – mainly local species, which hold the soil and prevents it from being washed away in the monsoon. Local grass cover can however only increase if free grazing of domestic animals is prevented by stall feeding. The next measure is to make '*nullah*' (small stream) plugs in the streams so that the water is held in the stream and does not rush down the hillside. In selected sites, several small check dams are built which together hold back larger amounts of water. All these measures constitute sound watershed management. It improves the water table and keeps the streams and *nullahs* flowing throughout the year.

Watershed Management Principles

This is a land management programme that looks at a region from the perspective of all its water-related issues. It can be used to manage a river from its source to its termination. Watershed management could also consider the management of a single valley as a unit, based on its small streams. Saving water from its local source by allowing it to percolate into the ground by *nullah* plugs and check dams instead of allowing it to run off rapidly along the surface during the monsoon, is a major aspect of good watershed management. This allows underground aquifers to fill so that groundwater is recharged. Deforestation is a major cause of poor water supply. Afforesting such degraded areas is an important aspect of watershed management.

Types of Watershed

Watershed is classified into a number of groups depending upon the mode of classification. The common modes of categorization are the size, drainage, shape and land use pattern. The categorization can also be based on the size of stream or river, the point of interception of the stream or the river and the drainage density and its distribution. The All India Soil and Land Use Surveys (AIS&LUS) of the Ministry of Agriculture, Government of India have developed a system for watershed delineation like water resource region, basin, catchment, sub-catchment and watershed. The usually accepted five levels of watershed delineation based on geographical area of the watershed are: macro watershed (> 50,000 ha), sub-watershed (10,000 to 50,000 ha), milli-watershed (1,000 to10,000 ha), micro watershed (100 to 1,000 ha) mini watershed (1–100 ha).

A watershed can be described as fan-shaped (near circular) or fen-shaped (elongated). Hydrologically, the shape of the watershed is important because it controls the time taken for the runoff to concentrate at the outlet. Watersheds may also be categorized as hill or plain watersheds, humid or arid watersheds, red soil watershed or black soil watershed based on criteria like soil, slope, climate etc. Depending on the land use pattern watershed can again be classified as highland watersheds, tribal, settlements and watersheds in areas of settled cultivation.

Components of Watershed Management

The three main components of watershed management are land management, water management and biomass management.

Figure 4.1 Components of Watershed Management

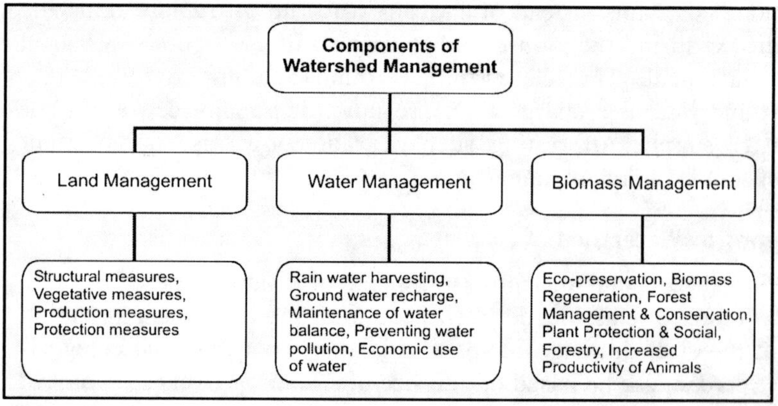

Land Management

Land characteristics like terrain, slope, formation, depth, texture, moisture, infiltration rate and soil capability are the major determinants of land management activities in a watershed. The broad category of land management interventions are: structural measures, vegetative measures, production measures and protection measures. Mechanical conservation measures may become necessary in watershed management in the initial stages. Structural measure include interventions like contour bunds, stone bunds, earthen bunds, graded bunds, compartmental bunds, contour terrace walls, contour trenches, bench terracing, broad based terraces, centripetal terraces, field bunds, channel walls, stream bank stabilization, check dams, etc. Watersheds may contain natural ecosystems like grasslands, wetlands, mangroves, marshes and water bodies. All these ecosystems have a specific role in nature. Vegetative measures include vegetative cover, plant cover, mulching, vegetative hedges, grass land management, vetiver fencing and agro-forestry. The production measures include interventions aimed at increasing the productivity of land like mixed cropping, strip cropping, cover cropping, crop rotations, cultivation of shrubs and herbs, contour cultivation conservation tillage, land levelling, use of improved verity of seeds, and horticulture. Protective measures like landslide control, gully plugging, runoff collection, etc., can also be adopted. Adoption of all the interventions mentioned above should be done strictly in accordance with the characteristics of the land taken for management.

Water Management

Water characteristics like inflows (precipitation, surface water and groundwater), water use (evaporation, evapotranspiration, irrigation and drinking water), outflows (surface water and groundwater) and storage (surface, groundwater and root zone) are the principal factors to be taken care of in sustainable water management. The broad interventions for water management are; rainwater harvesting, groundwater recharge, maintenance of water balance, preventing water pollution and economic use of water. Rainwater harvesting forms the major component of water management. The rainwater collected can be recharged into the ground. Rooftop water harvesting, diversion of perennial springs and streams into storage structures and farm ponds are the methods widely used for rainwater harvesting. Some simple and cost-effective rainwater harvesting structures are: percolation pits/tanks, recharge trenches/rain pits, recharge wells, Ferro-cement tanks, farm ponds, V ditch and bench terracing. Economic use of water and avoidance of affluence in use of water at individual and community levels may be the major concern for water management in the years to come. Watershed management can be attained through the following measures:

Conserving Water

Conserving water has become a prime environmental concern. Clean water is becoming increasingly scarce globally. With deforestation, surface runoff increases and the sub-soil water table drops as water has no time to seep slowly into the ground once vegetation is cleared. As many areas depend on wells, it has become necessary to go on making deeper and deeper wells. This adds to the cost and further depletes underground stores of water. This could take years to recharge even if the present rate of extraction is reduced which seems hardly possible in most situations. As deforestation and desertification spreads due to extensive changes in land use, the once perennial rivers are becoming increasingly seasonal. In many areas, the small streams run dry soon after the monsoon as the water table drops further and further below the surface. This adds serious problems caused by rapid surface flow of water during the rains, which leads to extensive floods with loss of life and property. When we waste water, we do not realize that it is affecting the lives of all of us in so many different ways. Water has to be equitably and fairly distributed so that household use, agriculture and industry all get a share of the water. Its overuse and misuse due to various activities that waste water or cause pollution has led to a serious shortage of potable drinking water. Thus, water conservation is linked closely with overall human well-being. Traditional systems of collecting water and using it optimally have been used in India for many generations. These have been forgotten in the recent past. Conserving water in multiple small percolation tanks and ponds was important in traditional forms of agriculture. Villages all over the country had one or more common 'talabs' or tanks from which people collected or used water carefully. As women had to carry water to their homes over long distances, this was a time consuming and labourious activity, thus the water could not be wasted. Many homes had a kitchen garden that was watered by the wastewater and in many remote areas, this practice is continued. Conservation of water was done in traditional homes through a conscious effort. During the British period, many dams were built across the country to supply water especially to growing urban areas. In the post-independence era, India's policy on water changed towards building large dams for expanding agriculture to support the green revolution. While this reduced the need to import food material and removed starvation in the country, the country began to see the effects of serious water shortages and problems related to its distribution. The newer forms of irrigated agriculture such as sugarcane and other water hungry cash crops required enormous quantities of water. Finally, however, such irrigated areas become waterlogged and unproductive. As excess water evaporates rapidly from the surface of heavily irrigated croplands, it pulls up sub-soil water along with salts to the surface of the soil. This leads to salinization by which the land becomes unproductive. Reducing the high salinity levels in soil is extremely expensive and frequently impossible. With all these ill-effects of the poorly conceived management of water at the national and local levels, there is a need to consider a new water policy for the country. Saving water in agriculture can be done through using various methods such as drip irrigation that supplies water to plants near its roots through a system of tubes. Small percolation tanks and rainwater harvesting can provide water for agriculture and domestic use.

Rainwater Harvesting

As the world faces serious water shortages, every drop of water we can use efficiently becomes of great value. One method is to manage rainwater in such a way that it is used at the source. If as much water as possible is collected and stored, this can be used after the rainy season is over. In many parts of the world, especially in very dry areas, this has been traditionally practised. However, the stored water has to be kept pollution free and clean so that it can be used as drinking water. Stored water can grow microscopic animals – algae and zooplankton. This can be pathogenic and cause infections. Thus, keeping the water uncontaminated is of great importance. Current technologies of rainwater harvesting require that all roof and terrace water passes down into a covered tank where it can be stored for use after the monsoon. This is most advantageous in arid areas where clean water is very scarce. However, there are practical difficulties such as constructing large storage tanks which are expensive. Another way of using rooftop rainwater harvesting is to collect it so that it percolates into the ground to recharge wells instead of flowing over the ground into rivers. Thus, by recharging groundwater harvested from rooftops, the water table rises and the surrounding wells retain water throughout the year.

Biomass Management

Biomass refers to the mass of living organisms in a particular environment, measured in terms of weight per unit area. At the present sphere of time, biomass degradation is high due to its overuse therefore, it is essential to conserve it. Biomass management can be done by eco-preservation, biomass regeneration, forest management and conservation, plant protection and social forestry, increased productivity of animals, income and employment generation programmes, better living standards for people, eco-friendly lifestyle of people, and formulation of a learning community.

People's Participation and Sustainability

People's participation and collective action are critical ingredients for watershed management. Sustainability, equity and participation are the three basic elements of participatory watershed management. Sustainability involves conservation and enhancement of the primary productivity of the ecosystem, the main components of which are land, water and biomass. Equity has to be seen in terms of creating an equitable access to livelihood resources for the watershed community. Participatory watershed management attempts at ensuring sustainability of the ecological, economic and social exchanges taking place in the watershed territory. This includes natural resource exchange, which is the conventional watershed management, and participatory watershed management additionally considers the economic, political and cultural exchanges. At this juncture, it may be advisable to understand the limitations of people's participation in any development project. Participation may lead to delayed start and slow progress in the initial stages of the programme. We may require more resources because in the participatory process we have to move along the path decided by the local people. Since participation is an empowering process where the people are empowered to make decisions, donors, governments and other players have to relinquish power and control. Relinquishing power and control is not an easy task for the bureaucracy. Increased expectation due to involvement of the local people may not always be accomplished. But, the advantages of people's participation are many and sound. Participation can ensure effective utilization of available resources. In real terms community participation means voluntary sharing by the users group their time, energy and money on the programme and adopt the recommended measures and practices on a sustained basis. Sustainability of participatory watershed management can be highlighted under three heads – socio-cultural indicators, economic indicators and the environmental/ecological indicators. The socio-cultural indicators deal with the social and cultural well-being of the watershed dwellers. This includes decision-making power of the community, empowerment of women, formation of farmer groups called self-help groups, change in ownership of land, improvement in quality of life and harmonious social life.

The economic indicators of sustainability consist of factors required for livelihood and economic well-being of the people, which consists of increase in income levels, availability of food and food security, improvement in standard of living, off-farm income to families, improvement in rural economy and improvements in credit and market supports. Environmental factors include tangible and non-tangible factors influencing the ecology of the community such as increase in the productive potential of resource base management of common property resources and improvement in bio-diversity.

Along with tremendous population growth and overexploitation of natural resources, need of watershed management increased manifold. Since, watershed management is a combined action for land, water and biomass management – life supporting layers, therefore a sustainable and timely watershed management can reduce the food scarcity and can restore the environment.

CHAPTER 5

Ecology

'Ecology' is the scientific study of the relationship of plants and animals to their physical and biological environment. The physical environment includes light and heat or solar radiation, moisture, wind, oxygen, carbon dioxide, nutrients in soil, water and atmosphere. The biological environment includes organisms of the same kind as well as other plants and animals. Because of the diverse approaches required to study organisms in their environment, ecology draws upon such fields as climatology, hydrology, oceanography, physics, chemistry, geology and soil analysis. To study the relationships between organisms, ecology also involves such disparate sciences as animal behaviour taxonomy, physiology, and mathematics. An increased public awareness of environmental problems has made 'ecology' a common term. Although, the field is a distinct scientific discipline, ecology does indeed contribute to the study and understanding of environmental problems. The term 'ecology' was coined by the German biologist Ernst Heinrich Haeckel in 1866. It is derived from the Greek *oikos,* which means household, sharing the same root word as economics. Thus, the term implies the study of the economy of nature. Modern ecology, in part, began with Charles Darwin. In developing his theory of evolution, Darwin stressed the adaptation of organisms to their environment through natural selection. Alexander von Humboldt also made important contribution in this field.

Biomes and Biosphere Reserves

All plants, animals and other organisms, as well as the physical environment in a particular area, are called biomes. A biome is characterized by its plant life, the types of which are determined by a location's climatic conditions, latitude and altitude. For example, northern coniferous forests exist in sub-arctic portions of North America and Asia, but further north, the conditions are simply too harsh and the season too brief for trees to grow. Instead of trees, the short vegetation of the tundra thrives in these areas. The same phenomenon occurs with altitude, as trees give way to short alpine vegetation in high mountainous regions. A biome is composed of many ecosystems – smaller communities of plants and animals and their habitats. While the boundaries of a biome are determined by climate, the boundaries of ecosystems are physical features, such as ridges or river banks that separate one community from another. The ecosystems of a particular biome tend to have plants with similar growth forms and animals with similar feeding habits. The major biomes in the world are as follows (Encarta, Encyclopedia, 2005):

Tropical Rainforest Biomes

Major terrestrial biomes include tropical rainforest, northern coniferous forest, tundra, desert, grassland, savanna and chaparral. Although not necessarily associated with marine environments, the term 'biome' is sometimes used by ecologists to describe marine life zones such as the littoral zone, found in shallow water; and the pelagic zone in the open ocean. The tropical rainforest is the most complex biome in the world. This biome is found at low elevations in the tropics where it is perpetually warm and wet.

The Tundra Biomes

The tundra is the treeless plain that lies north of the northern coniferous forests and on the Antarctic peninsula in the southern hemisphere. Trees cannot survive in this biome because of the cold temperatures, high winds and heavy snowfall, as well as the permafrost, a layer of permanently frozen sub-soil. Plant life tends to grow low to the ground. In the summer, large numbers of birds migrate to the tundra to feed on insects. Other animals found in this area include reindeer, wolves, fox, voles and lemmings. A similar biome, called the alpine biome, is found in high mountain areas around the world. Similar harsh conditions cause vegetation to grow low to the ground in alpine meadows.

The Desert Biomes

Desert biomes are characterized by less than 254 mm. of annual precipitation and high temperatures. To combat this lack of moisture, desert plants have developed water-conserving features, such as leaves that are light-coloured, small, thick, or waxy. Animals that live in the desert are often light-coloured, blend in well with their surroundings, and are usually more active at night to avoid the blazing heat during the day.

The Grassland Biomes

Grassland biomes are found on every continent except Antarctica, accounting for about one quarter of the earth's land surface. Typically found on flat or rolling terrain, grasslands tend to occur in the interior of continents where precipitation is lower. Periodic droughts occur in most grassland, accompanied by searing heat that scorches most vegetation in the area. Grasslands are covered with grasses, sedges, and other low-growing, perennial plants. Drought, fire and grazing by herbivores, such as bison and deer, restrict tree growth. Most grassland has been extensively cultivated and is now regions where major crops of wheat, corn and other grains are grown. Tropical savannas are expansive grasslands dotted with trees. The world's largest and best-known savanna is the African savanna, which covers much of the continent south of the Sahara desert. In the African savanna, herds of animals graze on the tall grass, and giraffes browse on the trees. Other tropical savannas are found in South America, India and Australia.

The Chaparral Biome

The chaparral biome is dominated by dense thickets of mostly small-leafed evergreen shrubs. It is characterized by hot, dry summers and cool, wet winters. This biome can be found in the foothills of California and Mediterranean climate regions. Chaparral plants have adapted to the frequent fires that result from lightning and dry conditions. The major difference between the two terms is that biomes include associated animal life.

Major biomes, however, go by the name of the dominant forms of plant life. Influenced by latitude, elevation, and associated moisture and temperature regimes, terrestrial biomes vary geographically from the tropics through the arctic and include various types of forest, grassland, shrub land and desert. These biomes also include their associated freshwater communities: streams, lakes, ponds and wetlands. Marine environments, also considered biomes by some ecologists, comprise the open ocean, littoral (shallow water) regions, benthic (bottom) regions, rocky shores, sandy shores, estuaries and associated tidal marshes.

Zone of air, soil and water that are capable of supporting life, traditionally thought of as a zone reaching about 10 km. into the atmosphere and down to the deepest ocean floor, is called 'biosphere'. Life in this zone depends primarily on the sun's energy through photosynthesis and on the circulation of heat and essential elements – most importantly the carbon cycle and the nitrogen cycle. A key feature of life on earth is the production of oxygen by plants and other photosynthetic organisms that take in carbon dioxide. In turn, animals and other organisms use oxygen in the process called 'cellular respiration' and release carbon dioxide. The known biosphere has been expanded by the discovery of organisms that use chemicals (chemosynthesis) rather than sunlight as an energy source to produce food. The earth's biosphere can be divided into three main habitable zones: the atmosphere, the hydrosphere and the lithosphere. Large-scale divisions of the biosphere into regions of different growth patterns are called 'biomes'. It has evolved over time. Primitive life is thought to have appeared on earth by about 3.8 billion years ago. Over time, the earth's atmosphere and ocean have undergone dramatic chemical changes. One of the most important events was the rise of oxygen produced by photosynthesis. Free oxygen in the atmosphere led to the formation of an ozone layer that protected the earth from deadly ultraviolet radiation, allowing organisms to live on land or at the surface of the ocean. The biosphere has remained sufficiently stable for hundreds of millions of years to sustain the evolution of today's life forms. However, the history of life includes mass extinctions that have greatly reduced the number of species of plants and animals. Activity by humans has had major effects on parts of the modern biosphere. Global warming, pollution, deforestation, desertification, over fishing, soil erosion, and other human-caused changes to the environment can pose a threat to the diversity of life on earth. Biosphere reserves have been established to help protect biodiversity.

National Parks and Preserves

National parks and preserves, unique public lands or bodies of water within a country, set aside by the government to protect ecosystems, plant and animal species, scenic landscapes, geologic formations, or historical or archaeological sites. National parks are managed primarily for public recreation, providing exceptional locations where visitors can view wildlife and enjoy the outdoors. Generally, these protected public lands are off-limits to hunting, livestock grazing, logging, mining, and other activities that exploit natural resources. Some parks commemorate significant historical events. National preserves, sometimes called nature or wildlife reserves, often are located within or near national parks. These lands are managed by national governments primarily for wildlife protection or scientific research instead of recreation, and provide 'living laboratories' in which scientists observe plant and animal species in their natural habitat.

The world's first national park, Yellowstone National Park, was created in the western United States in 1872 to reserve for public recreation 8,983 sq. km. of forest land. Since then countries around the world have established more than 4,000 national parks and preserves. Many national parks and preserves protect remote, unspoiled natural environments, while some protect islands of wilderness within heavily populated regions. From Arctic tundra to tropical grasslands, national parks and preserves protect a vast diversity of plant and animal life on land and in water throughout the world. Some of the world's major parks are: Arctic National

Park in Alaska, Serengeti National Park in Tanzania, Masai Mara National Reserve in Kenya, Tierra del Fuego National Park in Argentina, Rocky Mountain Park (now Banff National Park) in Canada, Palenque National Park in Mexico, Addo Elephant National Park and Mountain Zebra National Park in South Africa. In India, many national parks were established such as Kanha-Kilsi, Bandhav Garh, Corbett National Park, Rajaji National Park, Giri Wildlife Sanctuary and Rhino Sanctuary. Corbett National Park became India's first national park in 1936. Serving as a preserve for tigers, the 521 sq. km. park also protects leopards, several species of monkeys, crocodiles, myna birds, parakeets and woodpeckers. Tigers are more easily spotted in 392 sq. km. Ranthambore National Park (1980) than in any other park of India. The big cats can be spotted drinking from the park's lakes and even prowling about Ranthambore Fort, which was built when the park's land was a private hunting ground used by royalty. The Indian government is also working to preserve the Asiatic lion, a smaller cousin of the African lion. The Asiatic lion's only remaining natural habitat is in 259 sq. km., Gir National Park and Lion Sanctuary (1975), where the cat lives in thick forests of teak, ber, jamun and babul trees. Asiatic lions once roamed from the Middle East to India, but hunting reduced the wild population to fewer than 100 animals by the end of the 19th century. Protection programmes at Gir National Park helped the population rebound to nearly 300 by 1996.

Politics, economic recovery and prosperity, and emerging environmental science influenced the development of national parks and preserves on every continent following World War II. In the United States and Canada, peacetime prosperity made vacations possible for more people than ever before. More families also owned their own cars, and national parks quickly became popular destinations. Rebuilding from the devastation of the war, Europe's political and social leaders recognized a need to preserve and restore natural areas along with economic recovery. Calls from naturalists for the establishment of new national parks and protection of both plants and animals gained public support and resulted in positive government action. Movements for national independence from Europe's colonial powers arose across Africa and southern Asia, and included efforts to protect natural resources from exploitation by foreign governments. Today, these parks not only preserve plant and wildlife resources, but also ecosystems, natural beauty, geologic formations, recreation sites, archaeological treasures, and historic towns and villages. In 1971, the United Nations Educational, Scientific and Cultural Organization (UNESCO) and the New York Zoological Society, two of many conservation groups that promote the establishment and enhancement of national parks, successfully encouraged Ethiopia to create Bale Mountains National Park.

A leading organization in the creation and protection of national parks and preserves since the 1960s has been the World Wide Fund for Nature, known in North America as the World Wildlife Fund (WWF). One of WWF's major accomplishments occurred in 1973 when it helped the Government of India to organize Project Tiger, a cooperative programme for protecting tigers and their native habitat. Tigers are threatened throughout Asia by the black marketing of their body organs for folk medicines, and by deforestation and population growth that reduce both the tiger's habitat and its natural prey. Significant areas of several dozen tropical rainforests in Central and West Africa, Southeast Asia and Latin America have been protected as national parks or preserves through a WWF-sponsored campaign begun in 1975. Since 1985, WWF has invested more than $1.1 billion in more than 11,000 conservation projects around the world.

Mineral production and water pollution from mining threaten some preserves in India. In 1992 mining was permitted in Sariska Tiger Reserve (1958), which covers 492 sq. km. Logging, which destroys forest habitats and causes water pollution from soil erosion, and water projects for irrigation and flood control were allowed in 1,598 sq. km. Melghat Tiger Reserve (1985). Livestock grazing harms protected areas throughout the country. As the number of visitors to national parks throughout the world continues to rise, conflicts among tourists, scientists and wildlife increase. Cars, trucks and recreational vehicles jam popular destinations, polluting wilderness environments with vehicle exhaust emissions, litter and noise. Streams of hikers wear down park trails. Many visitors trample native plants by walking beyond designated trails, and some carry in non-native plants that threaten native vegetation.

Energy and Nutrients

Ecosystems function with energy flowing in one direction from the sun, and through nutrients, which are continuously recycled. Light energy is used by plants, which, by the process of photosynthesis, convert it to chemical energy in the form of carbohydrates and other carbon compounds. This energy is then transferred through the ecosystem by a series of steps that involve eating and being eaten, or what is called a *food web*. Each step in the transfer of energy involves several trophic, or feeding, levels: plants, herbivores (plant eaters), two or three levels of carnivores (meat eaters), and decomposers. Only a fraction of the energy fixed by plants follows this pathway, known as the *grazing food web*. Plant and animal matter not used in the grazing food chain, such as fallen leaves, twigs, roots, tree trunks, and the dead bodies of animals, support the *decomposer food web*. Bacteria, fungi and animals that feed on dead material become the energy source for higher trophic levels that tie into the grazing food web. In this way nature makes maximum use of energy originally fixed by plants. The number of trophic levels is limited in both types of food webs, because at each transfer a great deal of energy is lost and is no longer usable or transferable to the next trophic level. Thus, each trophic level contains less energy than the trophic level supporting it. For example, deer or caribou (herbivores) are more abundant than wolves (carnivores).

Energy flow fuels the biogeochemical or nutrient cycles. The cycling of nutrients begins with their release from organic matter by weathering and decomposition in a form that can be picked up by plants. Plants incorporate nutrients available in soil and water and store them in their tissues. The nutrients are transferred from one trophic level to another through the food web. Because most plants and animals go uneaten, nutrients contained in their tissues, after passing through the decomposer food web, are ultimately released by bacterial and fungal decomposition, a process that reduces complex organic compounds into simple inorganic compounds available for reuse by plants.

Figure 5.1 Food Web

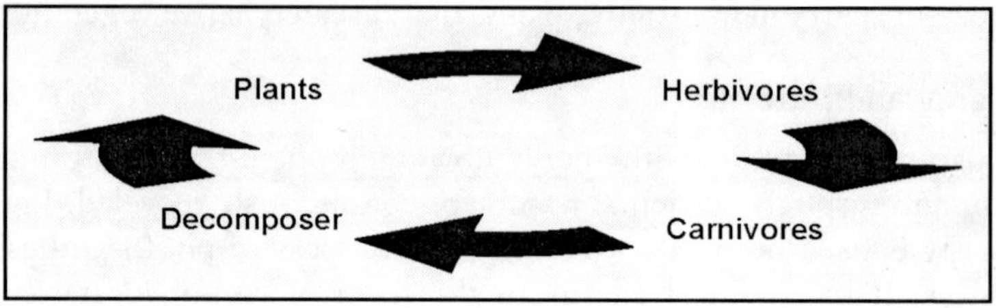

Imbalances

Within an ecosystem nutrients are cycled internally. But there are leakages or outputs, and these must be balanced by inputs, or the ecosystem will fail to function. Nutrient inputs to the system come from weathering of rocks, from windblown dust, and from precipitation, which can carry material to great distances. Varying quantities of nutrients are carried from terrestrial ecosystems by the movement of water and deposited in aquatic ecosystems and associated lowlands. Erosion and the harvesting of timber and crops remove considerable quantities of nutrients that must be replaced. The failure to do so results in an impoverishment of the ecosystem. This is why agricultural lands must be fertilized. If inputs of any nutrient greatly exceed outputs, the nutrient cycle in the ecosystem becomes stressed or overloaded, resulting in pollution. Pollution can be considered as an input of nutrients exceeding the capability of the ecosystem to process them. Nutrients eroded and leached from agricultural lands, along with sewage and industrial wastes accumulated from urban areas, all drain into streams, rivers, lakes, and estuaries. These pollutants destroy plants and animals that cannot tolerate their presence or the changed environmental conditions caused by them; at the same time they favour a few organisms more tolerant to changed conditions. Thus, precipitation filled with sulfur dioxide and oxides of nitrogen from industrial areas converts to weak sulfuric and nitric acids, known as acid rain, and falls on large areas of terrestrial and aquatic ecosystems. This upsets acid-base relations in some ecosystems, killing fish and aquatic invertebrates, and increasing soil acidity, which reduces forest growth in northern and other ecosystems that lack limestone to neutralize the acid.

Population and Communities

The functional units of an ecosystem are the population of organisms through which energy and nutrients move. A population is a group of interbreeding organisms of the same kind living in the same place at the same time. Groups of population within an ecosystem interact in various ways. These interdependent population of plants and animals make up the community, which encompasses the biotic portion of the ecosystem.

The community also has certain attributes such as dominance and species diversity. Dominance results when one or several species control the environmental conditions that influence associated species. In a forest, for example, the dominant species may be one or more species of trees, such as oak or spruce; in a marine community the dominant organisms frequently are animals such as mussels or oysters. Dominance can influence diversity of species in a community because diversity involves not only the number of species in a community, but also how numbers of individual species are apportioned. The physical nature of a community is evidenced by layering or stratification. In terrestrial communities, stratification is influenced by the growth form of the plants. Simple communities such as grasslands, with little vertical stratification, usually consist of two layers, the ground layer and the herbaceous layer. A forest has up to six layers: ground, herbaceous, low shrub, low tree and high shrub, lower canopy and upper canopy. These strata influence the physical environment and diversity of habitats for wildlife. Vertical stratification of life in aquatic communities, by contrast, is influenced mostly by physical conditions: depth, light, temperature, pressure, salinity, oxygen and carbon dioxide.

The community also provides the habitat – the place where particular plants or animals live. Within the habitat, organisms occupy different niches. A niche is the functional role of a species in a community – that is, its occupation, or how it earns its living. For example, the scarlet tanager lives in a deciduous forest habitat. Its niche, in part, is gleaning insects from the canopy foliage. The more a community is stratified, the more finely the habitat is divided into additional niches.

Population Growth Rate

Population have a birth rate (the number of young produced per unit of population per unit of time), a death rate (the number of deaths per unit of time), and a growth rate. While the major agent of population growth is births, the major agent of population loss is deaths. When births exceed deaths, a population increases; and when deaths exceed additions to a population, it decreases. When births equal deaths in a given population, its size remains the same, and it is said to be zero population growth. When introduced into a favourable environment with an abundance of resources, a small population may undergo geometric or exponential growth in the manner of compound interest. Many population experience exponential growth in the early stages of colonizing a habitat because they take over an underexploited niche or drive other population out of a profitable one. Those population that continue to grow exponentially, however, eventually reach the upper limits of the resources; they then decline sharply because of some catastrophic event such as starvation, disease, or competition from other species. In a general way, population of plants and animals that characteristically experience cycles of exponential growth are species that produce numerous young, provide little in the way of parental care, or produce an abundance of seeds having little food reserves. These species are usually short-lived, disperse rapidly and are able to colonize in harsh or disturbed environments. Such organisms are often called opportunistic species.

Other population tend to grow exponentially at first, and then logistically – that is, their growth slows as the population increases, and then levels off as the limits of their environment or carrying capacity are reached. Through various regulatory mechanisms, such population maintain something of equilibrium between their numbers and available resources. Animals exhibiting such population growth tend to produce fewer young but do provide them with parental care; the plants produce large seeds with considerable food reserves. These organisms are long-lived, have low dispersal rates, and are poor colonizers of disturbed habitats. They tend to respond to changes in population density (the number of organisms per unit area) through changes in birth and death rates rather than through dispersal. As the population approaches to the limit of resources, birth rates decline, and mortality of young and adults increases.

Community Interaction

Major influences on population growth involve various population interactions that tie the community together. These include competition, both within a species and among species; predation, including parasitism; and co-evolution, or adaptation.

Competition

When a shared resource is in short supply, organisms compete, and those that are more successful survive. Within some plant and animal populations, all individuals may share the resources in such a way that none obtains sufficient quantities to survive as adults or to reproduce. Among other plant and animal population, dominant individuals claim access to the scarce resources and others are excluded. Individual plants tend to claim and hold onto a site until they lose vigour or die. These prevent other individuals from surviving by controlling light, moisture, and nutrients in their immediate areas. Many animals have a highly developed social organization through which resources such as space, food, and mates are apportioned among dominant members of the population. Such competitive interactions may involve *social dominance,* in which the dominant individuals exclude subdominant individuals from the resource; or they may involve *territoriality,* in which the dominant individuals divide space into exclusive areas, which they defend. Sub-dominant or excluded individuals are forced to live in poorer habitats, do without the resource, or leave the area. Many of these animals succumb to starvation, exposure and predation. Competition among members of different species results in the division of resources in a community. Certain plants, for example, have roots that grow to different depths in the soil. Some have shallow roots that permit them to use moisture and nutrients near the surface. Others growing in the same place have deep roots that are able to exploit moisture and nutrients not available to surface-rooted plants.

Predation

One of the fundamental interactions is predation, or the consumption of one living organism, plant or animal, by another. While it serves to move energy and nutrients through the ecosystem, predation may also regulate population and promote natural selection by weeding the unfit from a population. Thus, a rabbit is a predator on grass, just as the fox is a predator on the rabbit. Predation on plants involves defoliation by grazers and the consumption of seeds and fruits. The abundance of plant predators, or herbivores, directly influences the growth and survival of the carnivores. Thus, predator-prey interactions at one feeding level influence the predator-prey relations at the next feeding level. In some communities, predators may so reduce population of prey species that a number of competing species can coexist in the same area because none is abundant enough to control the resource. When predators are reduced or removed, however, the dominant species tend to crowd out other competitors, thereby reducing species diversity.

Parasitism

Parasitism is closely related to predation, wherein two organisms live together, one drawing its nourishment at the expense of the other. Parasites, which are smaller than their hosts, include many viruses and bacteria. Because of this dependency relationship, parasites normally do not kill their hosts the way predators do. As a result, hosts and parasites generally coevolve a mutual tolerance, although parasites may regulate some host populations, lower their reproductive success, and modify behaviour.

Co-evolution

Co-evolution is the joint evolution of two unrelated species that have a close ecological relationship – that is, the evolution of one species depends in part on the evolution of the other. Co-evolution is also involved in predator-prey relations. Over a period of time, as predators evolve more efficient ways of capturing or consuming prey, the prey evolves ways to escape predation. Plants have acquired such defensive mechanisms as thorns, spines, hard seed-coats, and poisonous or ill-tasting sap that deter would-be consumers. Some herbivores are able to breach these defenses and attack the plant. Certain insects, such as the monarch butterfly, can incorporate poisonous substances found in food plants into their own tissues and use them as a defence against predators. Other animals avoid predators by assuming an appearance that blends them into the background or makes them appear part of the surroundings. The chameleon is a well-known example of this interaction. Some animals possessing obnoxious odours or poisons as a defence also have warning colourations, usually bright colours or patterns, that act as further warning signals to potential predators.

Another coevolutionary relationship is mutualism, in which two or more species depend on one another and cannot live outside such an association. An example of mutualism is mycorrhizae – an obligatory relationship between fungi and certain plant roots. In one group, called ectomycorrhizae, the fungi form a cap or mantle about the rootlets. The fungal hyphae (threads) invade the rootlet and grow between the cell walls as well as extending outward into the soil from the rootlet. The fungi, which include several common woodland mushrooms, depend on the tree for their energy source. In return the fungi aid the tree in obtaining nutrients from the soil and protect the rootlets of the tree from certain diseases. Without the mycorrhizae some groups of trees, such as conifers and oaks, cannot survive and grow. Conversely, the fungi cannot exist without the trees.

Succession and Climax Communities

Ecosystems are dynamic, in that the populations constituting them do not remain the same. This is reflected in the gradual changes of the vegetational community over time, known as succession. It begins with the colonization of a disturbed area, such as an abandoned crop field or a newly exposed lava flow, by species able to reach and to tolerate the environmental conditions present. Mostly these are opportunistic species that hold on to the site for a variable length of time. Being short-lived and poor competitors, they are eventually replaced by more competitive, longer-lived species such as shrubs, and then trees. In aquatic habitats, successional changes of this kind result largely from changes in the physical environment, such as the build-up of silt at the bottom of a pond. As the pond becomes shallower, it encourages the invasion of floating plants such as pond lilies and emergent plants such as cattails. The pace at which succession proceeds depends on the competitive abilities of the species involved; tolerance to the environmental conditions brought about by changes in vegetation; the interaction with animals, particularly the grazing herbivores; and fire. Eventually, the ecosystem arrives at a point called the *climax,* where further changes take place very slowly, and the site is dominated by long-lived, highly competitive species. As succession proceeds, however, the community becomes more stratified, enabling more species of animals to occupy the area. In time, animals characteristic of later stages of succession replace those found in earlier stages.

CHAPTER 6

Ecosystems

Ecosystem refers to organisms living in a particular environment, such as a forest or a coral reef, and the physical parts of the environment that affect them. The term 'ecosystem' was coined in 1935 by the British ecologist Sir Arthur George Tansley, who described natural systems in "constant interchange" among their living and non-living parts. The ecosystem concept fits into an ordered view of nature that was developed by scientists to simplify the study of the relationships between organisms and their physical environment, a field known as ecology. At the top of the hierarchy is the planet's entire living environment, known as the biosphere. Within this biosphere are several large categories of living communities known as biomes that are usually characterized by their dominant vegetation, such as grasslands, tropical forests, or deserts. The biomes are in turn made up of ecosystems. The living, or biotic, parts of an ecosystem, such as the plants, animals, and bacteria found in soil, are known as a community. The physical surroundings, or abiotic components, such as the minerals found in the soil, are known as the environment or habitat. An ecosystem is a collection of interdependent parts that function as a unit and involve inputs and outputs. The major parts of an ecosystem are the producers (green plants), the consumers (herbivores and carnivores), the decomposers (fungi and bacteria), and the non-living component, consisting of dead organic matter and nutrients in the soil and water. Inputs into the ecosystem are solar energy, water, oxygen, carbon dioxide, nitrogen, and other elements and compounds. Outputs from the ecosystem include water, oxygen, carbon dioxide, nutrient losses, and the heat released in cellular respiration, or heat of respiration. The major driving force is solar energy.

Classification of Ecosystems

Any given place may have several different ecosystems that vary in size and complexity (Encarta, Encyclopedia, 2005). A tropical island, for example, may have a rainforest ecosystem that covers hundreds of square miles, a mangrove swamp ecosystem along the coast, and an underwater coral reef ecosystem. No matter how the size or complexity of an ecosystem is characterized, all ecosystems exhibit a constant exchange of matter and energy between the biotic and abiotic community. Ecosystem components are so interconnected that a change in any one component of an ecosystem will cause subsequent changes throughout the system. The earth's zone of air, soil and water that are capable of supporting life, traditionally thought of as a zone reaching about 10 km. (about 6 miles) into the atmosphere and down to the deepest ocean floor is called 'biosphere'. Life in this zone depends primarily on the sun's energy through photosynthesis and on the circulation of heat and essential elements – most importantly the carbon cycle and the nitrogen cycle. A key feature of life on earth is the production of oxygen by plants and other photosynthetic organisms that take in carbon dioxide. In turn, animals and other organisms use oxygen in the process called cellular respiration and release carbon dioxide.

The known biosphere has been expanded by the discovery of organisms that use chemicals (chemosynthesis) rather than sunlight as an energy source to produce food. Extremophile organisms that live deep in the earth's crust, in scalding hydrothermal vents on the ocean floor, under ice, or in chemically hostile habitats suggest that some form of life may be possible in nearly any environment on earth in which liquid water and a source of energy is present. The existence of such organisms increases the possibility that some form of life might exist on other planets or moons in our solar system and elsewhere in the universe. The earth's biosphere can be divided into three main habitable zones: the atmosphere, the hydrosphere (water in the oceans, rivers and lakes) and the lithosphere (soil and the upper level of the crust). Large-scale divisions of the biosphere into regions of different growth patterns are called biomes.

The earth's biosphere has evolved over a long period of time. Primitive life is thought to have appeared on earth by about 3.8 billion years ago. The earliest organisms may have been similar to archaea – one-celled organisms that can live without oxygen or sunlight. Over time, the earth's atmosphere and oceans have undergone dramatic chemical changes. One of the most important events was the rise of oxygen produced by photosynthesis. Free oxygen in the atmosphere led to the formation of an ozone layer that protected the earth from deadly ultraviolet radiation, allowing organisms to live on land or at the surface of the ocean. Oxygen also allowed complex, multi-cellular life with high-energy metabolisms to evolve. The biosphere has remained sufficiently stable for hundreds of millions of years to sustain the evolution of today's life forms. However, the history of life includes mass extinctions that have greatly reduced the number of species of plants and animals. Activity by humans has had major effects on parts of the modern biosphere. Global warming, pollution, deforestation, desertification, overfishing, soil erosion, and other human-caused changes to the environment can pose a threat to the diversity of life on earth. Biosphere reserves have been established to help protect biodiversity.

Rainforest is the woodland characterized by lush vegetation and comparatively high temperature and rainfall throughout the year. Rainforests are the world's most biologically diverse ecosystems. Although, they account for less than 7 per cent of the land surface on earth, they contain more than 50 per cent – some scientists estimate as high as 90 per cent – of its plant and animal species. One hectare (about 2.5 acres) of tropical rainforest may contain more than 600 species of trees. By comparison, the forests of the United States and Canada altogether contain only around 700 tree species. Even more impressive are the number and diversity of animal species that call rainforests their home. One study found more species of ants living on a single rainforest stump than exist in all of the British Isles. Rainforests also play a critical role in global climate regulation by absorbing carbon dioxide – a greenhouse gas believed to be principally responsible for global warming. Plants naturally absorb carbon dioxide and give off oxygen in the process of photosynthesis, and tropical rainforests absorb more carbon dioxide than any other terrestrial ecosystem on earth. Global emissions of carbon dioxide have increased nearly 30 per cent in the last century. There is a general agreement among the scientific community that by absorbing some of the gas, rainforests play a vital role in lessening its impacts.

To be classified as a rainforest, a forest must have a closed canopy, in which the treetops, or crowns, touch each other, creating a shaded forest interior. In addition, temperature and rainfall must be high and relatively even throughout the year. Forests that meet these criteria are found flanking the equator in South and Central America, Asia, Africa and Australia. In South America, a vast, forested area of the Amazon river basin in Brazil and neighbouring countries is by far the largest rainforest in the world. It encompasses more than 3.5 million sq. km., about half of the total global rainforest covers. The larger of two large rainforests in Asia is centred along the Malay archipelago, including the islands of Borneo and Sumatra, the Malay peninsula, and the Republic of the Philippines. The other main rainforest in Asia is found primarily on the island of New Guinea and in northern Australia. In Africa, most of the rainforest is concentrated along the Atlantic coast and the Congo river basin. In regions of the northern and southern hemispheres, small areas of temperate rainforest are found along the coasts where rainfall and humidity are high and winters are mild. Such temperate rainforests are found in the Pacific northwest of the United States and southwestern Canada, for example. They are dominated by only a few tree species, however, making them differ dramatically from highly diverse tropical rainforests.

Mangrove is a common name applied to several kinds of tropical flowering plants that are members of three different families. Mangroves are trees or shrubs that have the common trait of growing in shallow and muddy salt water or brackish waters, especially along quiet shorelines and in estuaries. Typically, they produce tangled masses of arching roots that are exposed during low tides. Some mangrove roots extend above the water. These specialized vertical branches, called pneumatophores, act as aerating organs. The most common and most important species – the red mangrove – is so named because of its reddish wood. It has been exploited as a source of charcoal and of tannins, which are used in preparing leather. Its seeds are unusual in that they germinate while still on the tree, sprouting seedlings that grow up to 30 cm. long. The seedlings are cigar-shaped and heavier at the root end than at the leafy end. Upon falling, they tend to plant themselves in the mud below the parent tree. Red, white and button mangroves occur in southern Florida and in other tropical areas. The black mangrove is a member of the verbena family.

Scientific Classification

The red mangrove belongs to the family Rhizophoraceae and is classified as *Rhizophora mangle.* The white mangrove and button mangrove belong to the family Combretaceae. The black mangrove belongs to the family Verbenaceae and is classified as *Avicennia germinans.* Coral reefs are coastal and oceanic ecosystem renowned for its beautiful life forms and for providing one of the most biologically diverse habitats on earth. A reef is a ridge or outcrop of rock in the sea that comes close to the surface. A coral reef is a reef that has been built largely or entirely by corals, tiny animals that live together in colonies. Over hundreds or thousands of years the limestone skeletons of coral build up, with new corals growing on the skeletons of the dead ones. This physical structure, with its living surface of corals and other organisms, is a coral reef. The corals that build reefs are found only in warm tropical waters where sea temperatures rarely fall below 18°C. They thrive only in clear saltwater where bright sunlight can penetrate. This is because corals cannot exist without the symbiotic algae called zooxanthellae, that live in coral tissues and require sunlight for photosynthesis. Many reefs are found around islands, or quite far from land at the edge of continental shelves. Some of the most extensive areas of coral reefs are found in the Pacific Ocean, around the islands of Southeast Asia and off the coast of Australia. Large areas of coral reef are also found in the Indian Ocean and around the Red Sea and Arabian Sea. Many coral reefs also exist in the Caribbean Sea.

As their skeletons are made of limestone, corals leave clear and abundant fossils. Some of the earliest forms of life – mound- or pillar-shaped fossilized rocks known as stromatolites – resemble corals. The oldest stromatolites date back 3.5 billion years. They were laid down by some of the simplest organisms on earth – blue-green algae (also known as cyanobacteria). About 560 million years ago, limestone reefs built by bacteria, algae and sponges first appeared. The first complex corals, known as tabulate corals, are recorded at about 500 million years ago. The corals we know today first appeared about 220 million years ago, before the age of the dinosaurs. The first true coral reefs began to appear about 205 million to 210 million years ago. Since this time there have been several periods of major reef-building, but also periods when corals declined and did not build reefs. In all, shallow coral reefs occupy only about 2,84,000 sq. km., or less than 0.1 per cent of the world's oceans. Yet, in this tiny area, coral reefs house a quarter of all marine fish species. Reefs are highly productive resources for human beings. Hundreds of millions of people live within easy reach of coral reefs and rely on them to provide food. Coral reefs also help protect human settlements from large waves during storms.

How an Ecosystem Works?

The living portion of an ecosystem is best described in terms of feeding levels, known as *trophic levels*. Green plants make up the first trophic level and are known as primary producers. Plants are able to convert energy from the sun into food in a process known as photosynthesis. In the second trophic level, the primary consumers – known as herbivores – are animals and insects that obtain their energy solely by eating the green plants. The third trophic level is composed of the secondary consumers, flesh-eating or carnivorous animals that feed on herbivores. At the fourth level are the tertiary consumers, carnivores that feed on other carnivores. Finally, the fifth trophic level consists of the decomposers, organisms such as fungi and bacteria that break down dead or dying matter into nutrients that can be used again.

Some or all of these trophic levels combine to form what is known as a food web – the ecosystem's mechanism for circulating and recycling energy and materials. For example, in an aquatic ecosystem, algae and other aquatic plants use sunlight to produce energy in the form of carbohydrates. Primary consumers such as insects and small fish may feed on some of this plant matter, and are in turn eaten by secondary consumers, such as salmon. A brown bear may play the role of the tertiary consumer by catching and eating salmon. Bacteria and fungi may then feed upon and decompose the salmon carcass left behind by the bear, enabling the valuable non-living components of the ecosystem, such as chemical nutrients, to leach back into the soil and water, where they can be absorbed by the roots of plants. In this way nutrients and the energy that green plants derive from sunlight are efficiently transferred and recycled throughout the ecosystem. Photosynthesis is a process by which green plants and certain other organisms use the energy of light to convert carbon dioxide and water into the simple sugar glucose. In so doing, photosynthesis provides the basic energy source for virtually all organisms. An extremely important byproduct of photosynthesis is oxygen, on which most organisms depend. Photosynthesis occurs in green plants, seaweeds, algae, and certain bacteria. These organisms are veritable sugar factories, producing millions of new glucose molecules per second. Plants use much of this glucose, a carbohydrate, as an energy source to build leaves, flowers, fruits and seeds. They also convert glucose to cellulose, the structural material used in their cell walls. Most plants produce more glucose than they use, however, and they store it in the form of starch and other carbohydrates in roots, stems and leaves. The plants can then draw on these reserves for extra energy or building materials. Each year, photosynthesizing organisms produce about 170 billion metric tons of extra carbohydrates, about 30 metric tons for every person on earth.

Photosynthesis has far-reaching implications. Like plants, humans and other animals depend on glucose as an energy source, but they are unable to produce it on their own and must rely ultimately on the glucose produced by plants. Moreover, the oxygen humans and other animals breathe is the oxygen released during photosynthesis. Humans are also dependent on ancient products of photosynthesis, known as fossil fuels, for supplying most of our modern industrial energy. These fossil fuels, including natural gas, coal and petroleum, are composed of a complex mix of hydrocarbons – the remains of organisms that relied on photosynthesis millions of years ago. Thus, virtually, all life on earth, directly or indirectly, depends on photosynthesis as a source of food, energy and oxygen, making it one of the most important biochemical processes known. Food web is a set of interconnected food chains by which energy and materials circulate within an ecosystem. The food web is divided into two broad categories: the grazing web, which typically begins with green plants, algae, or photosynthesizing plankton, and the detrital web, which begins with organic debris. These webs are made up of individual food chains. In a grazing web, materials typically pass from plants to plant eaters (herbivores) to flesh eaters (carnivores). In a detrital web, materials pass from plant and animal matter to bacteria and fungi (decomposers), then to detrital feeders (detritivores), and then to their predators (carnivores). Generally, many interconnections exist within food webs. For example, the fungi that decompose matter in a detrital web may sprout mushrooms that are consumed by squirrels, mice, and deer in a grazing web. Robins are omnivores, that is, consumers of both plants and animals, and thus are in both detrital and grazing webs. Robins typically feed on earthworms, which are detritivores that feed upon decaying leaves.

Through these series of steps of eating and being eaten, energy flows from one trophic level to another. Green plants or other photosynthesizing organisms use light energy from the sun to manufacture carbohydrates for their own needs. Most of this chemical energy is processed in metabolism and dissipated as heat in respiration. Plants convert the remaining energy to biomass, both above ground as woody and herbaceous tissue, and below ground as roots. Ultimately, this material, which is stored energy, is transferred to the second trophic level, which comprises grazing herbivores, decomposers and detrital feeders. Most of the energy assimilated at the second trophic level is again lost as heat in respiration; a fraction becomes new biomass. Organisms in each trophic level pass on as biomass much less energy than they receive. Thus, the more steps between producer and final consumer, the less energy remains available. Seldom are there more than four links, or five levels, in a food web. Eventually, all energy flowing through the trophic levels is dissipated as heat. The process whereby energy loses its capacity to do work is called entropy.

Ecosystem Management

Humans benefit from these smooth-functioning ecosystems in many ways. Healthy forests, streams and wetlands contribute to clean air and clean water by trapping fast-moving air and water, enabling impurities to settle out or be converted to harmless compounds by plants or soil. The diversity of organisms, or biodiversity, in an ecosystem provides essential foods, medicines, and other materials. But, as human population increases and its encroachment on natural habitats expands, humans are having detrimental effects on the very ecosystems on which they depend. The survival of natural ecosystems around the world is threatened by many human activities: bulldozing wetlands and clear-cutting forests – the systematic cutting of all trees in a specific area – to make room for new housing and agricultural land; damming rivers to harness the energy for electricity and water for irrigation; and polluting the air, soil and water.

Many organizations and government agencies have adopted a new approach to managing natural resources – naturally occurring materials that have economic or cultural value, such as commercial fisheries, timber and water – in order to prevent their catastrophic depletion. This strategy, known as ecosystem management, treats resources as interdependent ecosystems rather than simply commodities to be extracted. Using advances in the study of ecology to protect the biodiversity of an ecosystem, ecosystem management encourages practices that enable humans to obtain necessary resources using methods that protect the whole ecosystem. As regional economic prosperity may be linked to ecosystem health, the needs of the human community are also considered.

Ecosystem management often requires special measures to protect threatened or endangered species that play key roles in the ecosystem. In the commercial shrimp trawling industry, for example, ecosystem management techniques protect loggerhead sea turtles. In the last thirty years, population of loggerhead turtles on the southeastern coasts of the United States has been declining at alarming rates due to beach development and the ensuing erosion, bright lights and traffic, which makes it nearly impossible for female turtles to build nests on beaches. At sea, loggerheads are threatened by oil spills and plastic debris, offshore dredging, injury from boat propellers, and getting caught in fishing nets and equipment. In 1970, the species was listed as threatened under the Endangered Species Act.

When scientists learnt that commercial shrimp trawling nets were trapping and killing between 5,000 and 50,000 loggerhead sea turtles a year, they developed a large metal grid called a Turtle Excluder Device (TED) that fits into the trawl net, preventing 97 per cent of trawl-related loggerhead turtle deaths while only minimally reducing the commercial shrimp harvest. In 1992, the National Marine Fisheries Service (NMFS), implemented regulations requiring commercial shrimp trawlers to use TEDs, effectively balancing the commercial demand for shrimp with the health and vitality of the loggerhead sea turtle population.

CHAPTER 7

Biodiversity

Biodiversity or biological diversity implies a sum of all the different species of animals, plants, fungi, and microbial organisms living on earth and the variety of habitats in which they live. Scientists estimate that more than 100 million different species inhabit the earth. Each species is adapted to its unique niche in the environment, from the peaks of mountains to the depths of deep-sea hydrothermal vents, and from polar ice caps to tropical rainforests. Biodiversity underlies everything from food production to medical research. Humans the world over use at least 40,000 species of plants and animals on a daily basis. Many people around the world still depend on wild species for some or all of their food, shelter and clothing needs. All of our domesticated plants and animals came from wild-living ancestral species. The array of living organisms found in a particular environment, together with the physical and environmental factors that affect them, is called an ecosystem. Healthy ecosystems are vital to life, as they regulate many of the chemical and climatic systems that make available clean air and water and plentiful oxygen. Forests, for example, regulate the amount of carbon dioxide in the air, produce oxygen as a byproduct of photosynthesis (the process by which plants convert energy from sunlight into carbohydrate energy), and control rainfall and soil erosion. Ecosystems, in turn, depend on the continued health and vitality of the individual organisms that compose them. Removing just one species from an ecosystem can prevent the ecosystem from operating optimally.

Measuring Biological Diversity

The greatest value of biodiversity is yet unknown. Scientists have discovered and named only 1.75 million species – less than 20 per cent of those estimated to exist and of those identified, only a fraction has been examined for potential medicinal, agricultural, or industrial value. Much of the earth's great biodiversity is rapidly disappearing, even before we know what is missing. Most biologists agree that life on earth is now faced with the most severe extinction episode since the event that drove the dinosaurs to extinction 65 million years ago. Species of plants, animals, fungi, and microscopic organisms such as bacteria are being lost at alarming rates – so many, in fact, that biologists estimate that three species go extinct every hour. Scientists around the world are cataloging and studying global biodiversity in hopes that they might better understand it, or at least slow the rate of loss. Everywhere, there is life; there is more than one distinct type of organism. Even a drop of seawater offers a multitude of different microscopic plants, animals, and less complex life forms. The rich diversity of the living world is connected in two distinct ways. First, different types of organisms live side by side in complex ecological networks of interdependency, each relying on the others that share its habitat for nutrients and energy. Second, all life on earth is connected with an evolutionary tree of life. At the bottom of the tree are the common ancestor from which all living things descended – a single-celled microbe that lived more than 3.5 billion years ago – and in its uppermost branches are gorillas, chimpanzees, orangutans, and our own species – *Homo sapiens*.

Ecological diversity is the intricate network of different species present in local ecosystems and the dynamic interplay between them. An ecosystem consists of organisms from many different species living together in a region that are connected by the flow of energy, nutrients, and matter that occurs as the organisms of different species interact with one another. The ultimate source of energy in nearly all ecosystems is the sun. The sun's radiant energy is converted to chemical energy by plants. This energy flows through the systems when animals eat the plants and then are eaten, in turn, by other animals. Fungi derive energy by decomposing organisms, releasing nutrients back into the soil as they do so. An ecosystem, then, is a collection of living components – microbes, plants, animals and fungi – and non-living components – climate and chemicals – that are connected by energy flow. Removing just one species from an ecosystem damages the flow of energy of that system. For instance, in the late 19th and early 20th centuries, sea otters were hunted to near extinction in many kelp forests off the coast of the Pacific, northwest of the United States and western Canada, causing the entire ecosystem to suffer. Otters eat sea urchins, small, spiny organisms that share their habitat. When the otters disappeared, the sea urchin population exploded and started to destroy the vast beds of kelp. Without the kelp, other species that lived in the ecosystem, including many species of fish and snails and other invertebrates, began to decline in number. Efforts to restore sea otter populations brought the kelp communities back to near normal in the late 20th century.

Measuring ecological diversity is difficult because each of the earth's ecosystems merges into the ecosystems around it. A lake, for example, might have a distinct shoreline, but the plants fringing its edges are quite different from the aquatic plants in the middle of the lake or the trees and shrubs surrounding the lake. Beavers may live in the lake, but they construct dams from trees that grow in adjacent ecosystems. Nutrients flow into the lake via streams and rivers beyond the lake's ecosystem. Every species on earth is related to every other species in a pattern every bit as complex as the patterns of energy flow within an ecosystem. In evolutionary diversity, the connection is not energy flow, but rather genetic connections that unite species. The more closely related any two species are, the more genetic information they will share, and the more similar they will appear. An ever-widening circle of evolutionary relatedness embraces every species on earth.

An organism's closest relatives are members of its own species – that is, other organisms with which it has the potential to mate and produce offspring. Members of a species share genes, the bits of biochemical information that determine, in part, how the animals look, behave and live. One eastern gray squirrel, for example, shares the vast majority of its genes with other eastern gray squirrels, whether they live in the same area or are separated by thousands of miles. Members of a species also share complex mating behaviours that enable them to recognize each other as potential mates. When a female eastern gray squirrel is ready to mate, she exudes a scent that attracts male eastern gray squirrels. Mating and sharing a common supply of genes unite a species. For virtually every species there is a similar and closely related species in an adjacent habitat. West of the Rocky mountains, one finds western gray squirrels instead of eastern gray squirrels. Although western gray squirrels are more similar to than different from their eastern counterparts, these animals do not share a common mating behaviour with eastern gray squirrels. Even when brought into close proximity, eastern and western gray squirrels do not mate, and so constitute two distinct species.

Each species has other, more remotely related species, which share a more general set of characteristics. Gray squirrels, chipmunks, marmots, and prairie dogs all belong to the squirrel family because they share a number of features, such as tooth number and shape, and details of skull and muscle anatomy. All of these animals are rodents, a large group of more distantly related animals who share similar chisel-like incisor teeth that grow continuously. All rodents are related to a broader group mammals. Mammals have hair, raise their young ones on milk, and have three bones in the middle ear. All mammals, in turn, are more distantly related to other animals with backbones, or vertebrates. All these organisms are animals but share a common cell structure with plants, fungi, and some microbes. Finally, all living organisms share a common molecule, ribonucleic acid (RNA), and most also have deoxyribonucleic acid (DNA). These molecules direct the production of proteins – molecules responsible for the structure and function of virtually all living cells. This is the evolutionary chain of life. All species are descended from a single common ancestor. From that ancient single-celled microbe, all inherited RNA. As time goes by, species diverge and develop their own peculiar attributes, thus, making their own contribution to biodiversity.

Global Biodiversity Crises

Most biologists accept the estimate of American evolutionary biologist Edward O. Wilson that the earth is losing approximately 27,000 species per year. This estimate is based primarily on the rate of disappearance of ecosystems, especially tropical forests and grasslands, and our knowledge of the species that live in such systems. We can measure the rate of loss of tropical rainforests, for example, by analysing satellite photographs of continents from different periods that show rates and amounts of habitat destruction – and from these measurements calculate the approximate number of species being lost each year. This extraordinary rate of extinction has occurred only five times before in the history of complex life on earth. Mass extinctions of the geological past were caused by catastrophic physical disasters, such as climate changes or meteorite impacts, which destroyed and disrupted ecosystems around the globe. In the mass extinction, which occurred more than 65 million years ago, the Earth was shrouded in a cloud of atmospheric dust – the result of meteorite impact or widespread volcanic activity. The resulting environmental disruption caused the demise of 76 per cent of all species alive at the time, including the dinosaurs. Today's extinction is likewise primarily caused by ecosystem disturbance – but this time the destroying force is not the physical environment, but rather humankind. The human transformation of the earth's surface threatens to be every bit as destructive as any of the past cataclysmic physical disasters.

Human Impact

The underlying cause of biodiversity loss is the explosion in human population, now at 7 billion (2011), but expected to double again by the year 2050. The human population already consumes nearly half of all the food, crops, medicines, and other useful items produced by the earth's organisms, and more than 1 billion people on earth lack adequate supplies of fresh water. But the problem is not sheer numbers of people alone: The unequal distribution and consumption of resources and other forms of wealth on the planet must also be considered. According to some estimates, an average middle-class American consumes an amazing 30 times what a person living in a developing nation consumes. Thus, the impact of the 270 million American people must be multiplied by 30 to derive an accurate comparative estimate of the impact such industrialized nations have on the world's ecosystems.

The single greatest threat to global biodiversity is the human destruction of natural habitats. Since the invention of agriculture about 10,000 years ago, the human population has increased from approximately 5 million to a full 6 billion people. During that time, but especially in the past several centuries, humans have radically transformed the face of the planet earth. The conversion of forests, grasslands, and wetlands for agricultural purposes, coupled with the multiplication and growth of urban centres and the building of dams and canals, highways and railways, has physically altered ecosystems to the point that extinction of species has reached its current alarming pace. In addition, overexploitation of the world's natural resources, such as fisheries and forests, has greatly outstripped the rate at which these systems can recover. For example, 12 of the 13 largest oceanic fisheries are severely depleted. Modern fishing techniques, such as using huge fishing nets and bottom vacuuming techniques, remove everything in their paths – including tons of fish and invertebrates of no commercial use. These victims, as well as porpoises and seals that are also hauled in as accidental catches, are permanently removed from their populations, significantly altering the ecosystems in which they live. As human populations have grown, people have spread out to the four corners of the earth. In the process, they have introduced nonnative species, which have created ecological nightmares, disrupting local ecosystems and, in many cases, directly driving native species extinct. For example, the brown tree snake was introduced to the island of Guam, probably as a stowaway on visiting military cargo ships after World War II (1939–1945). The snake devastated the native bird population, driving over half a dozen native species of birds to extinction – simply because the native birds had not been exposed to this type of predator and did not recognize the danger posed by these snakes.

Conserving Biodiversity

As the scope and significance of biodiversity loss can be better understood now positive steps to stem the tide of the extinction have been proposed and, to some extent, adopted. Several nations have enacted laws protecting endangered wildlife. An international treaty known as the Convention on International Trade in Endangered Species of Wild Fauna and Flora (CITES) went into effect in 1975 to outlaw the trade of endangered animals and animal parts. In the United States, the Endangered Species Act (ESA) was enacted in 1973 to protect endangered or threatened species and their habitats. The Convention on Biological Diversity, held in Rio de Janeiro, Brazil, in 1992 and ratified by more than 160 countries, obligates governments to take action to protect plant and animal species. In the last three decades, focus has shifted away from the preservation of individual species to the protection of large tracts of habitats linked by corridors that enable animals to move between the habitats. Thus, the movement to save species, for example, the spotted owl of the Pacific northwest, has become an effort to protect vast tracts of old-growth timber.

Promising as these approaches may be, conservation efforts will never succeed in the long run if the local economic needs of people living in and near threatened ecosystems are not taken into account. This is particularly true in developing countries, where much of the world's remaining undisturbed land is located. At the end of the 20th century, international organizations such as the World Bank and the World Wildlife Fund (WWF) launched a movement for all countries in the developing world to set aside 10 per cent of their forests in protected areas. But many communities living near these protected areas have relied on the rain forest for food and firewood for thousands of years. Left with few economic alternatives, these communities may be left without enough food to eat. To address this problem, the burgeoning field of conservation biology emphasizes interaction with the people directly impacted by conservation measures. Conservation biologists encourage such people to develop sustainable economic alternatives to destructive harvesting and land use. One alternative is harvesting and selling renewable rainforest products, such as vegetable ivory seeds from palms, known as tagua nuts, and Brazil nuts. Where protection measures permit, rainforest communities may undertake sustainable rainforest logging operations, in which carefully selected trees are extracted in a way that has minimal impact on the forest ecosystem. Still other communities are exploring medicinal plants for drug development as ways to strengthen and diversify their economies.

Conservation biologists also work with established industries to develop practices that ensure the health and the sustainability of the resources on which they depend. For example, conservation biologists work with fishermen to determine how many fish the fishermen can harvest without damaging the population and the ecosystem as a whole. The same principles are applied to the harvesting of trees, plants, animals, and other natural resources. Preserving biodiversity also takes place at the molecular level in the conservation of genetic diversity. All around the world efforts are being made to collect and preserve endangered organisms' DNA – the molecule that contains their genes. These collections, or *gene banks*, may consist of frozen samples of blood or tissue, or in some cases, they may consist of live organisms. Biologists use gene banks to broaden the gene pool of a species, increasing the likelihood that it will adapt to meet the environmental challenges that confront it. Many zoos, aquariums and botanical gardens work together to carefully maintain the genetic diversity in captive population of endangered animals and plants, such as the giant panda, the orangutan, or the rosy periwinkle. Captive animals are bred with wild population, or occasionally released in hopes that they will breed freely with members of the wild population, thus increasing its genetic diversity. These gene banks are also an essential resource to replenish the genetic diversity of crops, enabling plant breeders and bio-engineers to strengthen their stocks against diseases and changing climatic conditions (Encarta Encyclopedia, 2005).

CHAPTER 8

Environmental Pollution

With the advancement in science and technology all over the world, environmental pollution increased manifold and, consequently, large-scale environmental degradation took place. Environmental pollution in the forms of air, water, noise and soil is widespread. These forms of pollution are classified as follows (Encarta Encyclopedia, 2005):

Air Pollution

Air pollution adds harmful substances to the atmosphere resulting in damage to the environment, human health and quality of life. One of many forms of pollution – air pollution – occurs inside homes, schools and offices; in cities; across continents; and even globally. Air pollution makes people sick – it causes breathing problems and promotes cancer – and it harms plants, animals and the ecosystems in which they live. Some air pollutants return to earth in the form of acid rain and snow, which corrode statues and buildings, damage crops and forests, and make lakes and streams unsuitable for fish and other plant and animal life. Pollution is changing earth's atmosphere so that it lets in more harmful radiation from the sun. At the same time, our polluted atmosphere is becoming a better insulator, preventing heat from escaping back into space and leading to a rise in global average temperatures. Scientists predict that the temperature increase, referred to as global warming, will affect world food supply, alter sea level, make weather more extreme, and increase the spread of tropical diseases.

Major Air Pollutant Sources

Most air pollution comes from one human activity: burning fossil fuels natural gas, coal, and oil – to power industrial processes and motor vehicles. Among the harmful chemical compounds this burning puts into the atmosphere are carbon dioxide, carbon monoxide, nitrogen oxides, sulfur dioxide, and tiny solid particles – including lead from gasoline additives – called particulates. Between 1900 and 1970, motor vehicle use has rapidly expanded, and emissions of nitrogen oxides, some of the most damaging pollutants in vehicle exhaust, increased upto 690 per cent. When fuels are incompletely burned, various chemicals called volatile organic chemicals (VOCs) also enter the air. Pollutants also come from other sources. For instance, decomposing garbage in landfills and solid waste disposal sites emits methane gas, and many household products give off VOCs. Some of these pollutants also come from natural sources. For example, forest fires emit particulates and VOCs into the atmosphere. Ultra-fine dust particles dislodged by soil erosion, when water and weather loosen layers of soil, increase air-borne particulate levels. Volcanoes spew out sulfur dioxide and large amounts of pulverized lava rock known as volcanic ash. A big volcanic eruption can darken the sky over a wide region and affect the earth's entire atmosphere. The 1991 eruption of Mount Pinatubo in the Philippines, for example, dumped enough volcanic ash into the upper atmosphere to lower global temperatures for the next two years. Unlike pollutants from human activity, however, naturally occurring pollutants tend to remain in the atmosphere for a short time and do not lead to permanent atmospheric change. Once in the atmosphere, pollutants often undergo chemical reactions that produce additional harmful compounds. Air pollution is subject to weather patterns that can trap it in valleys or blow it across the globe to damage pristine environments far from the original sources.

Local and Regional Pollution

Local and regional pollution take place in the lowest layer of the atmosphere – the troposphere – which at its widest extends from earth's surface to about 16 km. The troposphere is the region in which most weather occurs. If the load of pollutants added to the troposphere were equally distributed, the pollutants would be spread over vast areas and the air pollution might almost escape our notice. Pollution sources tend to be concentrated, however, especially in cities. In the weather phenomenon known as thermal inversion, a layer of cooler air is trapped near the ground by a layer of warmer air above. When this occurs, normal air mixing almost ceases and pollutants are trapped in the lower layer. Local topography, or the shape of the land, can worsen this effect – an area ringed by mountains, for example, can become a pollution trap.

Box 8.1 Smog and Acid Precipitation

Smog is intense local pollution usually trapped by a thermal inversion. Before the age of the automobile, most smog came from burning coal. Burning gasoline in motor vehicles is the main source of smog in most regions today. Smog contains ozone, a form of oxygen gas made up of molecules with three oxygen atoms rather than the normal two. Ozone in the lower atmosphere is a poison. It damages vegetation, kills trees, irritates lung tissues, and attacks rubber. In cities around the world, smog acids are eroding precious artifacts, including the Parthenon temple in Athens, Greece, and the Taj Mahal in Agra, India. Oxides of nitrogen and sulfur dioxide pollute places far from the points where they are released into the air. Carried by winds in the troposphere, they can reach distant regions where they descend in acid form, usually as rain or snow. In extreme cases, smog can lead to mass illness and death, mainly from carbon monoxide poisoning. In 1948 in the steel-mill town of Donora, Pennsylvania, intense local smog killed 19 people. In 1952, in London, about 4,000 people died in one of the notorious smog events known as London Fogs; in 1962 another 700 Londoners died. The worst of such accidents occurred in 1984 in Bhopal, India, when methyl isocyanate released from an American-owned factory during a thermal inversion caused more than 3,800 deaths.

Global Scale Pollution

Air pollution can expand beyond a regional area to cause global effects. The stratosphere is the layer of the atmosphere between 16 km. (10 miles) and 50 km. (30 miles) above sea level. It is rich in ozone, the same molecule that acts as a pollutant when found at lower levels of the atmosphere in urban smog. Up to the stratospheric level, however, ozone forms a protective layer that serves a vital function: It absorbs the wavelength of solar radiation known as ultraviolet-B (UV-B). UV-B damages deoxyribonucleic acid (DNA), the genetic molecule found in every living cell, increasing the risk of such problems as cancer in humans. Because of its protective function, the ozone layer is essential to life on earth.

Box 8.2 Ozone Depletion

Several pollutants attack the ozone layer. Chief among them is the class of chemicals known as chlorofluorocarbons (CFCs), formerly used as refrigerants (notably in air conditioners), as agents in several manufacturing processes, and as propellants in spray cans. CFC molecules are virtually indestructible until they reach the stratosphere. Here, intense ultraviolet radiation breaks the CFC molecules apart, releasing the chlorine atoms they contain. These chlorine atoms begin reacting with ozone, breaking it down into ordinary oxygen molecules that do not absorb UV-B. The chlorine acts as a catalyst – that is, it takes part in several chemical reactions – yet at the end emerges unchanged and able to react again. A single chlorine atom can destroy up to 1,00,000 ozone molecules in the stratosphere. Other pollutants, including nitrous oxide from fertilizers and the pesticide methyl bromide, also attack atmospheric ozone. Scientists are finding that under this assault the protective ozone layer in the stratosphere is thinning. In the Antarctic region, it vanishes almost entirely for a few weeks every year. Although CFC use has been greatly reduced in recent years and will soon be prohibited worldwide, CFC molecules already released into the lower atmosphere will be making their way to the stratosphere for decades, and further ozone loss is expected. As a result, experts anticipate an increase in skin cancers, more cataracts (clouding of the lens of the eye), and reduced yields of some food crops.

Indoor Air Pollution

Pollution is perhaps most harmful at an often unrecognized site inside the homes and buildings where we spend most of our time. Indoor pollutants include tobacco smoke; radon, an invisible radioactive gas that enters homes from the ground in some regions; and chemicals released from synthetic carpets and furniture, pesticides, and household cleaners. When disturbed, asbestos, a non-flammable material once commonly used in insulation, sheds air-borne fibres that can produce a lung disease called asbestosis. Pollutants may accumulate to reach much higher levels than they do outside, where natural air currents disperse them. Indoor air levels of many pollutants may be 2 to 5 times, and occasionally more than 100 times, higher than outdoor levels. These levels of indoor air pollutants are especially harmful because people spend as much as 90 per cent of their time living, working, and playing indoors. Inefficient or improperly vented heaters are particularly dangerous.

Water Pollutants

Water pollution is contamination of streams, lakes, underground water, bays and oceans by substances harmful to living things. Water is necessary to life on earth. All organisms contain it, live in it and drink it. Plants and animals require water that is moderately pure, and they cannot survive if their water is loaded with toxic chemicals or harmful micro-organisms. If severe, water pollution can kill large numbers of fish, birds, and other animals, in some cases killing all members of a species in an affected area. Pollution makes streams, lakes, and coastal waters unpleasant to look at, to smell, and to swim in. Fish and shellfish harvested from polluted waters may be unsafe to eat. People who ingest polluted water can become ill, and, with prolonged exposure, may develop cancers or bear children with birth defects.

Types of Water Pollutants

The major water pollutants are chemical, biological or physical materials that degrade water quality. Pollutants can be classified into eight categories, each of which presents its own set of hazards.

Petroleum Product

Oil and chemicals derived from oil are used for fuel, lubrication, plastics manufacturing, and many other purposes. These petroleum products get into water, mainly by means of accidental spills from ships, tanker trucks, pipelines, and leaky underground storage tanks. Many petroleum products are poisonous if ingested by animals, and spilled oil damages the feathers of birds or the fur of animals, often causing death. In addition, spilled oil may be contaminated with other harmful substances, such as polychlorinated biphenyls (PCBs).

Pesticides and Herbicides

Chemicals used to kill unwanted animals and plants, for instance on farms or in suburban yards, may be collected by rainwater runoff and carried into streams, especially if these substances are applied too lavishly. Some of these chemicals are biodegradable and quickly decay into harmless or less harmful forms, while others are non-biodegradable and remain dangerous for a long time. When animals consume plants that have been treated with certain nonbiodegradable chemicals, such as chlordane and dichlorodiphenyltrichloroethane (DDT), these chemicals are absorbed into the tissues or organs of the animals. When other animals feed on these contaminated animals, the chemicals are passed up the food chain. With each step up the food chain, the concentration of the pollutant increases. This process is called biomagnification. In one study, DDT levels in ospreys (a family of fish-eating birds) were found to be 10 to 50 times higher than in the fish that they ate, 600 times the level in the plankton that the fish ate, and 10 million times higher than in the water. Animals at the top of food chains may, as a result of these chemical concentrations, suffer cancers, reproductive problems, and death. Many drinking water supplies are contaminated with pesticides from widespread agricultural use.

Heavy Metals

Heavy metals such as copper, lead, mercury and selenium get into water from many sources, including industries, automobile exhaust, mines, and even natural soil. Like pesticides, heavy metals become more concentrated as animals feed on plants and are consumed in turn by other animals. When they reach high levels in the body, heavy metals can be immediately poisonous, or can result in long-term health problems similar to those caused by pesticides and herbicides. For example, cadmium in fertilizer derived from sewage sludge can be absorbed by crops. If these crops are eaten by humans in sufficient amounts, the metal can cause diarrhea and, over time, liver and kidney damage. Lead can get into water from lead pipes and solder in older water systems; children exposed to lead in water can suffer mental retardation.

Hazardous Waste

Hazardous wastes are chemical wastes that are toxic, reactive, corrosive and ignitable. If improperly treated or stored, hazardous wastes can pollute water supplies. In 1969 the Cuyahoga river in Cleveland, Ohio, was so polluted with hazardous wastes that it caught fire and burned. PCBs, a class of chemicals once widely used in electrical equipment such as transformers, can get into the environment through oil spills and can reach toxic levels as organisms eat one another.

Excess Organic Matters

Fertilizers and other nutrients used to promote plant growth on farms and in gardens may find their way into water. At first, these nutrients encourage the growth of plants and algae in water. However, when the plant matter and algae die and settle underwater, micro-organisms decompose them. In the process of decomposition, these micro-organisms consume oxygen that is dissolved in the water. Oxygen levels in the water may drop to such dangerously low levels that oxygen-dependent animals in the water, such as fish, die. This process of depleting oxygen to deadly levels is called eutrophication.

Sediments

Sediment, soil particles carried to a streambed, lake, or ocean, can also be a pollutant if it is present in large enough amounts. Soil erosion produced by the removal of soil-trapping trees near waterways, or carried by rainwater and floodwater from croplands, strip mines, and roads, can damage a stream or lake by introducing too much nutrient matter. This leads to eutrophication. Sedimentation can also cover streambed gravel in which many fish, such as salmon and trout, lay their eggs.

Infectious Organism

A 1994 study by the Centres for Disease Control and Prevention (CDC) estimated that about 9,00,000 people get sick annually in the United States because of organisms in their drinking water, and around 900 people die. Many disease-causing organisms that are present in small numbers in most natural waters are considered pollutants when found in drinking water. Such parasites as *Giardia lamblia* and *Cryptosporidium parvum* occasionally turn up in urban water supplies. These parasites can cause illness, especially in people who are very old or very young, and in people who are already suffering from other diseases. In 1993 an outbreak of *Cryptosporidium* in the water supply of Milwaukee, Wisconsin, sickened more than 4,00,000 people and killed more than 100.

Sources of Water Pollutants

Water pollutants result from many human activities. Pollutants from industrial sources may pour out from the outfall pipes of factories or may leak from pipelines and underground storage tanks. Polluted water may flow from mines where the water has leached through mineral-rich rocks or has been contaminated by the chemicals used in processing the ores. Cities and other residential communities contribute mostly sewage, with traces of household chemicals mixed in. Sometimes industries discharge pollutants into city sewers, increasing the variety of pollutants in municipal areas. Pollutants from such agricultural sources as farms, pastures, feedlots and ranches contribute animal wastes, agricultural chemicals, and sediment from erosion. The oceans, vast as they are, are not invulnerable to pollution. Pollutants reach the sea from adjacent shorelines, from ships, and from offshore oil platforms. Sewage and food waste discarded from ships on the open sea do little harm, but plastics thrown overboard can kill birds or marine animals by entangling them, choking them, or blocking their digestive tracts if swallowed. Oil spills often occur through accidents, such as the wrecks of the tanker *Amoco Cadiz* off the French coast in 1978 and the *Exxon Valdez* in Alaska in 1992. Routine and deliberate discharges, when tanks are flushed out with seawater, also add a lot of oil to the oceans. Offshore oil platforms also produce spills. The largest oil spill ever was the result of an act of war. During the Gulf War of 1991, Iraqi forces destroyed eight tankers and onshore terminals in Kuwait, releasing a record 910 million litres. An oil spill has its worst effects when the oil slick encounters a shoreline. Oil in coastal waters kills tidepool life and harms birds and marine mammals by causing feathers and fur to lose their natural waterproof quality, which causes the animals to drown or die of cold. Additionally, these animals can become sick or poisoned when they swallow the oil while preening. Water pollution can also be caused by other types of pollution. For example, sulfur dioxide from a power plant's chimney begins as air pollution. The polluted air mixes with atmospheric moisture to produce air-borne sulfuric acid, which falls to the earth as acid rain. In turn, the acid rain can be carried into a stream or lake, becoming a form of water pollution that can harm or even eliminate wildlife. Similarly, the garbage in a landfill can create water pollution if rainwater percolating through the garbage absorbs toxins before it sinks into the soil and contaminates the underlying groundwater.

Solid Waste

Solid wastes are unwanted solid materials such as garbage, paper, plastics and other synthetic materials, metals and wood. Billions of tons of solid waste are thrown out annually. The United States alone produces about 200 million metric tons of municipal solid waste each year. A typical American generates an average of 2 kg. of solid waste each day. Cities in economically developed countries produce far more solid waste per capita than those in developing countries. Moreover, waste from developed countries typically contains a high percentage of synthetic materials that take longer to decompose than the primarily biodegradable waste materials of developing countries. Areas where wastes are buried, called landfills, are the cheapest and most common disposal method for solid wastes worldwide. But landfills quickly become overfilled and may contaminate air, soil and water. Incineration, or burning, of waste reduces the volume of solid waste but produces dense ashen wastes that often contain dangerous concentrations of hazardous materials such as heavy metals and toxic compounds. Composting, using natural biological processes to speed the decomposition of organic wastes, is an effective strategy for dealing with organic garbage and produces a material that can be used as a natural fertilizer. Recycling, extracting and reusing certain waste materials, has become an important part of municipal solid waste strategies in developed countries. According to the EPA, more than one-fourth of the municipal solid waste produced in the United States is now recycled or composted. Recycling also plays a significant, informal role in solid waste management for many Asian countries, such as India, where organized waste-pickers comb streets and dumps for items such as plastics, which they use or resell. Expanding recycling programmes worldwide can help reduce solid waste pollution, but the key to solving severe solid waste problems lies in reducing the amount of waste generated. Waste prevention, or source reduction, such as altering the way products are designed or manufactured to make them easier to reuse, reduces the high costs associated with environmental pollution.

Hazardous Waste

Industrial Waste

Hazardous wastes is generated by nearly every industry; those industries that themselves generate few hazardous waste nonetheless use products from hazardous waste generating industries. For example, in the computer software industry, writing software generates little hazardous waste, but the manufacture of computers involves many industrial processes. Making a computer circuit board generates spent electroplating baths that contain metal salts, and the production of computer chips uses acids, other caustic chemicals and solvents. Other hazardous waste is generated in the manufacture of fibre optics and copper wire used in electronic transmission, as well as magnetic disks, paper for technical manuals, photographs for packaging and publicity, and trucks for transportation of the finished product.

Agricultural Waste

Industry is not alone in generating hazardous wastes. Agriculture produces such waste as pesticides and herbicides and the materials used in their application. Fluoride waste implies by-products of phosphate fertilizer production. Even soluble nitrates from manure may dissolve into groundwater and contaminate drinking water wells; high levels of nitrates may cause health problems.

Household Waste

Household sources of hazardous waste include toxic paints, flammable solvents, caustic cleaners, toxic batteries, pesticides, drugs, and mercury from broken fever thermometers. Local waste disposal systems may refuse these items. If they are accepted, careful monitoring may be required to make sure soil or groundwater is not contaminated. The householder may be asked to recycle or dispose of these items separately. Renovations of older homes may cause toxic lead paint to flake off from walls. Insulation material on furnace pipes may contain asbestos particles, which can break off and hang suspended in air; and when inhaled, they can cause lung disease and cancer.

Medical Waste

Hospitals use special care in disposing of wastes contaminated with blood and tissue, separating the hazardous wastes from ordinary waste. Hospitals and doctors' offices must be especially careful with needles, scalpels and glassware, called "sharps". Pharmacies discard outdated and unused drugs; testing laboratories dispose of chemical waste. Medicine also makes use of significant amounts of radioactive isotopes for diagnosis and treatment, and these substances must be tracked and disposed of carefully.

Effects of Hazardous Waste

Hazardous waste may pollute soil, air, surface water or underground water. Pollution of soil may affect people who live on it, plants that put roots into it, and animals that move over it. Sludge from municipal sewage disposal may contain toxic elements if industrial waste is mixed with domestic sewage. If the sludge is used as a fertilizer, these elements may contaminate fields. Toxic substances that do not break down or bind tightly to the soil may be taken up by growing plants; the toxic substances may later appear in animals that eat crops grown there and possibly in people who do so.

Air may become contaminated by direct emission of hazardous waste. Evaporation of toxic solvents from paints and cleaning agents is a common problem. The air above hazardous waste may become dangerously contaminated by escaping gas, as can occur in houses built on mine tailings or old dump sites. Basements of homes built over uranium mine tailings often contain high levels of radioactive radon gas escaping from the radioactivity below. River and lake pollution, if it is toxic enough, may kill animal and plant life immediately, or it may injure slowly. For example, fluoride concentrates in teeth and bone, and too much fluoride in water may cause dental and bone problems. Compounds such as DDT, PCBs and dioxins are more soluble in fats than in water and therefore tend to build up in the fats within plants and animals. These substances may be present in very low concentrations in water but accumulate to higher concentrations within algae and insects, and build up to even higher levels in fish. Birds or people that feed on these fish are then exposed to very high levels of hazardous substances. In birds, these substances can interfere with egg production and bone formation.

Hazardous Waste Control

The best way to eliminate hazardous waste is not to generate them in the first place. For example, improvements have been made in the production of integrated circuits. The toxic chlorinated hydrocarbons commonly used in the 1970s were replaced in the 1980s by less toxic glycol ethers and in the 1990s by low-toxicity esters and alcohols.

Recycling

Recycling is the recovery or reuse of usable materials from waste. About 5 per cent of hazardous waste in the United States is recycled as solvents; a similar amount is recovered as metals. For example, approximately 15 per cent of sulfuric acid is recycled in chemical manufacturing. In the past, most sulfur used for sulfuric acid production was mined; now, the amount of sulfur recovered from smelters, refineries and manufacturers is more than double the produced by mining. In the United States, the practice of using industrial waste, which often contain hazardous elements, as ingredients in commercial fertilizers, is encouraged as a means of recycling hazardous waste. The safety of this practice has recently been called into question, however, and some states are starting to regulate it.

Treatment

Waste may be made less hazardous by physical, chemical or biological treatment. Nearly 10 per cent of hazardous waste in the United States is treated with water; another 11 per cent undergoes other treatment. For example, sodium hydroxide has been used to treat acid waste at integrated-circuit plants. Some newer plants now treat hydrofluoric acid waste with lime, producing relatively harmless calcium fluoride, the mineral fluorite. Sulfuric acid waste, if not recycled, can be treated with ammonia waste from the same plant, forming ammonium sulfate, a fertilizer. Incineration has been used since human beings learned to control fire. It is the preferred method of handling infectious medical waste. However, it should not be used for wastes that contain toxic heavy metals or chlorinated hydrocarbons: when burned, old painted surfaces can release lead or arsenic into the air, whereas chlorinated hydrocarbons produce hydrochloric acid and dioxins. Solids left over from incineration may have to be disposed of as hazardous waste. About 6 per cent of hazardous waste in the United States is incinerated, and another 11 per cent is burned along with fuel.

Disposal

Surface impoundment (placing liquid or semi-liquid waste in unlined pits) keeps waste in long-term storage, but it is not considered a method of final disposal. About 8 per cent of hazardous waste is injected into deep wells; 21 per cent enters landfills (large, unlined pits into which solid waste is placed) as its ultimate resting place. In the Unites States abandoned and particularly serious waste sites may qualify as "Superfund" sites, eligible for clean up with government funding under legislation passed in 1980. In 1993, of about 38,000 hazardous waste sites inventoried by the Environmental Protection Agency (EPA), 1,407 sites were listed on or proposed for the National Priority List (NPL) for waste clean up. In 1995 the EPA estimated that 73 million people lived within four miles of a Superfund site in the United States. Before 1995, 3,300 emergency removals-urgent clean ups of hazardous waste because of the immediate hazard they present – were conducted. The serious problem of underground plumes of hazardous materials leaving the original disposal sites has only partial solutions at this time. The typical method of handling this problem is the drilling of wells around a plume's perimeter. Hazardous materials are then removed from some wells, and water may be injected into other wells to produce a barrier to the plume's motion. Drilling wells and monitoring holes near a toxic site carries risks; a plume originally confined between strata may penetrate vertically through a drilled hole and escape confinement. A recent method of treatment for shallow plumes of chlorinated solvents depends on their chemical reactivity. A trench is dug around the leaking waste site and filled with a mixture of soil and powdered iron. The iron then reacts with the chlorinated solvents, turning them into simple hydrocarbons, which are less hazardous.

Legislative History

Hazardous waste has a broad definition in common use, but according to United States law, hazardous waste has a narrow, specific meaning. In 1965, recognizing the need to segregate hazardous materials from general solid wastes – unwanted solid or semisolid substances – to prevent soil and groundwater contamination, the Congress of the United States passed the Solid Waste Disposal Act (SWDA), to address problems of solid waste disposal and landfills. In 1976 the Toxic Substances Control Act (TSCA), regulating the use and management – including disposal – of PCBs and other toxic substances, was passed. PCBs had been widely used in the insulation of transformers and capacitors in electric power systems and had been discovered to cause cancer. Also in 1976 the Resource Conservation and Recovery Act (RCRA) were passed, regulating the generation, transportation and management of hazardous wastes. In the 1984 reauthorization of RCRA, Congress added the Hazardous and Solid Waste Amendments (HSWA). In 1979 at Love Canal, near Niagara Falls, New York, and in 1983 at Times Beach, Missouri, neighbourhoods were declared uninhabitable because they were discovered to have been dump sites for hazardous wastes. To control and clean up past hazardous-waste disposal sites, Congress passed the Comprehensive Environmental Response, Compensation, and Liability Act (CERCLA) of 1980, which came to be known as the "Superfund," and its 1986 reauthorization, the Superfund Amendments and Reauthorization Act (SARA). Under RCRA, hazardous waste is a type of solid waste. Solid waste is defined quite broadly and may include solids, sludge, liquids, or gases. Wastes that are not considered solid waste include domestic sewage or wastes that pass through a publicly owned treatment works, industrial discharges subject to the Clean Water Act (1977, amended 1992 and 1995) or Clean Air Act (1970, amended 1977 and 1990), irrigation water, nuclear materials, mine wastes remaining in the ground, recycled sulfuric acid, and some other recycled materials. Radioactive waste is specifically controlled by the Atomic Energy Act (AEA) of 1954, and medical waste by the Medical Waste Tracking Act of 1988. Some of the materials excluded from the RCRA definition of hazardous wastes are agricultural wastes used as fertilizers; overburden (material overlying mineral deposits) returned to mines; chromium wastes; materials contaminated with petroleum from the clean up of underground tanks, such as those used at gas stations; wastes from the burning of coal or other fossil fuels; wastes from oil, gas, and geothermal exploration and wells; and household waste. Worldwide, about 400 million metric tons of hazardous waste is generated each year. In 1989 the Basel Convention on the Control of Transboundary Movements of Hazardous Wastes and Their Disposal was adopted at a meeting convened by the United Nations Environmental Programme and attended by 116 countries. The Convention requires reduction of hazardous waste and their movements across borders between countries. As of January 1996, 97 countries had ratified the Convention; the United States was not yet among them. The import of hazardous waste has been prohibited by about 90 countries.

Noise or Sound Pollution

Noise pollution or sound pollution means exposure of people or animals to levels of sound that are annoying, stressful or damaging to the ears. Although loud and frightening sounds are part of nature, only in recent centuries has much of the world become urban, industrial and chronically noisy. Sound intensity is measured in units called decibels. The decibel scale is logarithmic and climbs steeply: an increase of about three decibels is a doubling of sound volume. In the wilderness, a typical sound level would be 35 decibels. Speech runs 65 to 70 decibels; heavy traffic generates 90 decibels. By 140 decibels, sound becomes painful to the human ear, but ill-effects, including hearing loss, set in at much lower levels. Most noise pollution comes from machines, especially automobiles, trucks and aircrafts. Construction equipment, farm machines and the din of machinery inside factories can be dangerously loud. Some home appliances, shop tools, lawnmowers and leaf blowers can also be noisy, as are guns, firecrackers and some toys. Even music, when played at very high volume, particularly through personal headphones, is as damaging to the ears as a roaring chain saw. Even the oceans are increasingly noisy. The engines of ships, especially large vessels and supertankers, produce a lot of sound, and traffic is increasing. Offshore oil exploration and drilling are also noisy. And, more recently, very loud, low-frequency sonar signals are being sent through the seas to detect submarines and to gain information about water temperatures and global warming.

Impact of Noise Pollution

The most significant health problem caused by noise pollution is hearing loss. Any noise appreciably louder than talking can damage the delicate hair cells in the cochlea – the structure in the inner ear that converts sound waves into auditory nerve signals. The initial damage to the cochlea may be temporary, but with repeated exposure, the damage becomes permanent. Loud noise deafens quickly-extremely loud sounds, such as gunshots at close range, can cause immediate hearing loss. But even sound levels of only 85 decibels can cause some hearing loss after prolonged exposure. Ten million Americans have some hearing loss due partly or wholly to exposure to loud sounds, and 20 million are at risk. In addition to deafness, many people with damaged ears are afflicted with tinnitus, or ringing in the ears. Most hearing loss occurs in workplaces, where workers may be unable to avoid unhealthy noise, and where exposure may continue for years. Factory workers, construction workers, farmers, military personnel, police officers, fire-fighters and musicians all have reason to be concerned about their occupational exposure to noise. Even at levels below those that cause hearing loss, noise pollution produces problems. Noise makes conversation difficult, interferes with some kinds of work and disturbs sleep. As a source of stress, it can promote high blood pressure and other cardiovascular problems, as well as nervous disorders. According to the National Institute of Health, 65 million Americans are exposed to noise levels that can hamper their work or disrupt their sleep, and 25 million risk health problems due to noise. Noise also puts stress on domestic animals and wildlife. In remote areas, helicopters and military aircraft often frighten animals. Aircraft noise in Alaska, for example, has been shown to reduce the survival rate of caribou calves. There is concern that increasing noise levels in the oceans may confuse the natural sonar that whales use to navigate, communicate and locate food.

Controlling Noise Pollution

Noise pollution is not a necessary price to pay for living in an industrial society. Much can be done to reduce the severity of the problem. For example, vehicles and other machines can be built to produce less noise. Four-cycle engines can replace much noisier two-cycle engines in such products as lawnmowers, motorboats, and jet skis. Labels that indicate the noise levels of appliances and tools can help consumers avoid noisy products and choose quieter alternatives. Even after noise is generated, steps can be taken to reduce human exposure to it. At homes or in offices, insulation of walls and double-glazing of windows can muffle sound from traffic, neighbours, and other sources from the outside world. Sound walls along highways can shield nearby neighbourhoods from traffic noise. Individuals should protect themselves with earplugs or muff like ear protectors, particularly when noise levels exceed 85 decibels.

In the industrialized nations, governments have laws and policies to counter noise pollution. In the United States, for example, at least six federal agencies are involved in controlling noise pollution. Since 1969 the Federal Aviation Administration (FAA) has monitored and controlled noise from airplanes. The agency requires that new aircraft meet specified noise standards and that old ones be retrofitted or retired. Local airport authorities, with FAA approval, reduce the impacts of noise pollution by routing flights over water or unpopulated areas on takeoff and landing, and by limiting traffic at night. The FAA also encourages airports and local governments to take steps on the ground, such as constructing sound barriers, insulating buildings, and restricting residential development in noisy areas. In extreme cases, airports have relocated people living under flight paths. The Occupational Safety and Health Administration (OSHA) is charged with reducing noise in workplaces. Under OSHA regulations, no exposure above 115 decibels is permitted, exposure up to 115 decibels is limited to 15 minutes for an 8–hour shift, and average noise levels above 85 decibels are regulated. OSHA requires employers to measure noise levels, to muffle extremely noisy equipment, to provide ear-protection gear if necessary, and to offer regular hearing tests to workers who are regularly exposed to high sound levels. The Bureau of Mine Safety has comparable rules to protect miners. The Department of Housing and Urban Development and the Veterans Administration require noise-proofing in dwellings whose mortgages they finance. The Department of Defense even has noise standards for certain military situations.

In 1972 Congress passed a Noise Control Act establishing an Office of Noise Abatement and Control in the Environmental Protection Agency. The office conducted research, coordinated the work of other agencies, and directly set noise standards for trucks, motorcycles, air compressors, truck-mounted garbage compactors and railroads. More standards would have followed, but in 1981 Congress cut off funding for this effort. Some state and many local governments work to reduce noise pollution. State and local building codes include noise insulation requirements, and land use planning is used to keep noise sources away from housing and offices. Local ordinances can ban the use of some equipment, such as leaf blowers, or limit use to certain times of day. In the last 30 years, the United States and other countries have expended considerable efforts to control noise pollution. Most vehicles and many other noise producers are quieter than they used to be. On the other hand, there are more noise-making machines than ever, operating most of the time. In the United States, most of the rules governing noise pollution were established at least two decades ago, and critics are calling for new, stronger measures, as well as for better enforcement of the old ones. What is true of most other kinds of pollution is also true of noise. Our best efforts against it tend merely to keep matters from getting dramatically worse.

Thermal Pollution

Water is often drawn from rivers, lakes, or the ocean for use as a coolant in factories and power plants. The water is usually returned to the source warmer than when it was taken. Even small temperature changes in a body of water can drive away the fish and other species that were originally present, and attract other species in place of them. Thermal pollution can accelerate biological processes in plants and animals or deplete oxygen levels in water. The result may be fish and other wildlife deaths near the discharge source. Thermal pollution can also be caused by the removal of trees and vegetation that shade and cool streams.

Major Sources of Thermal Pollution

The major sources of thermal pollution are electric power plants and industrial factories. In most electric power plants, heat is produced when coal, oil, or natural gas is burned or nuclear fuels undergo fission to release huge amounts of energy. This heat turns water to steam, which in turn spins turbines to produce electricity. After doing its work, the spent steam must be cooled and condensed back into water. To condense the steam, cool water is brought into the plant and circulated next to the hot steam. In this process, the water used for cooling warms 5 to 10°C, after which it may be dumped back into the lake, river, or ocean from which it came. Similarly, factories contribute to thermal pollution when they dump water used to cool their machinery. The second type of thermal pollution is much more widespread. Streams and small lakes are naturally kept cool by trees and other tall plants that block sunlight. People often remove this shading vegetation in order to harvest the wood in the trees, to make room for crops, or to construct buildings, roads, and other structures. Left unshaded, the water warms by as much as 10°C. In a similar manner, grazing sheep and cattle can strip stream sides of low vegetation, including young trees. Even the removal of vegetation far away from a stream or lake can contribute to thermal pollution by speeding up the erosion of soil into the water, making it muddy. Muddy water absorbs more energy from the sun than clear water does, resulting in further heating. Finally, water running off of artificial surfaces, such as streets, parking lots, and roofs, is warmer than water running off vegetated land and, thus, contributes to thermal pollution.

Impact of Thermal Pollution

All plant and animal species that live in water are adapted to temperatures within a certain range. When water in an area warms more than they can tolerate, species that cannot move, such as rooted plants and shellfish, die. Species that can move, such as fish, leave the area in search of cooler conditions, and they die if they cannot find them. Typically, other species, often less desirable, move into the area to fill the vacancy. In general, cold waters are better habitat for plants and animals than warm ones because cold waters contain more dissolved oxygen. Many freshwater fish species that are valued for sport and food, especially trout and salmon, do poorly in warm water. Some organisms do thrive in warm water, often with undesirable effects. Algae and other plants grow more rapidly in warm water than in cold, but they also die more rapidly; the bacteria that decompose their dead tissue use up oxygen, further reducing the amount available for animals. The dead and decaying algae make the water look, taste, and smell unpleasant.

Controlling Thermal Pollution

Thermal pollution from power plants and factories is relatively easy to control. Instead of discharging heated water into lakes and streams, power plants and factories can pass the heated water through cooling towers or cooling ponds, where evaporation cools the water before it is discharged. Alternatively, power plants can be designed or refitted to be more efficient and to produce less waste heat in the first place. In a process called *cogeneration,* the excess heat energy from generating electricity can be used in another manufacturing process that needs such energy. Where homes or other buildings are located near industrial plants, waste hot water can be used for heating – an arrangement often found in Scandinavian towns and cities, and proposed for use in China. In the United States, the problem of industrial thermal pollution was first addressed in 1970, when Congress gave the Atomic Energy Commission authority to regulate thermal pollution from nuclear power plants. In 1972, the comprehensive Clean Water Act instructed the Environmental Protection Agency to issue regulations to clean up all hot water discharges from all power plants, nuclear or conventional. Since then, thermal pollution from power plants has not been a major issue in the United States.

To prevent thermal pollution due to devegetation, the prescription is simple: do not devegetate. Landowners can leave strips of trees and vegetation along streams and shorelines. Grazing livestock can be kept away from streamsides by fencing. All efforts to control erosion also have the effect of keeping water clearer and, thus, cooler. As a practical matter, however, thermal pollution from devegetation is quite hard to control because it is caused by the cumulative effect of many people's actions, most of which are individually minor. Regulations focus on a few of the most important threats. Grazing management plans, for instance, are intended to counter thermal pollution and other problems on lands owned by the federal government. In the United States, regulations governing logging on both public and private lands supposedly protect streamsides, though enforcement is often lax. Elsewhere, streamside protection is largely up to private landowners, encouraged and aided by such advisory organizations as the federal Natural Resources Conservation Service and cooperative Resource Conservation Districts.

Soil Pollution

Soil is a mixture of mineral, plant, and animal materials that form during a long process that may take thousands of years. It is necessary for most plant growth and is essential for all agricultural production. Soil pollution is caused by toxic chemical compounds, salts, pathogens, or radioactive materials that can affect plant and animal life. Unhealthy soil management methods have seriously degraded soil quality, caused soil pollution, and enhanced erosion. Treating the soil with chemical fertilizers, pesticides, and fungicides interferes with the natural processes occurring within the soil and destroys useful organisms such as bacteria, fungi, and other micro-organisms. For instance, strawberry farmers in California fumigate the soil with methyl bromide to destroy organisms that may harm young strawberry plants. This process indiscriminately kills even beneficial micro-organisms and leaves the soil sterile and dependent upon fertilizer to support plant growth. This results in heavy fertilizer use and increases polluted runoff into lakes and streams. Improper irrigation practices in areas with poorly drained soil may result in salt deposits that inhibit plant growth and may lead to crop failure. In 2000 BC, the ancient Sumerian cities of the southern Tigris-Euphrates Valley in Mesopotamia depended on thriving agriculture. By 1500 BC, these cities had collapsed largely because of crop failure due to high soil salinity. The same soil pollution problem exists today in the Indus Valley in Pakistan, the Nile Valley in Egypt, and the Imperial Valley in California.

CHAPTER 9

Environmental Hazards

The dictionary meaning of 'hazard' is something that is potentially very dangerous. In the beginning, hazards were believed to be natural only, in which volcanic eruptions, earthquakes and tectonic movements were highly dangerous and unavoidable, coupled with atmospheric hazards, which include storms of various types, droughts, floods, snow, hail, lightning and natural fires. But, the present dimensions of hazards have also evolved with the human-induced activities, which include large-scale depletion of forests, emission of major greenhouse gases into the atmosphere, and high rate of developmental activities. Among natural hazards, tropical hurricanes are the worst and the most widespread natural hazard, causing damage not only directly by wind action but also by flooding. Floods and droughts are often closely linked in both space and time, with many parts of the tropics having alternating wet and dry seasons. Increased incidence of drought has caused major civilizations to collapse in the past and currently much of Africa is becoming drier, particularly on the fringes of the Sahara, where the problem is exacerbated by poor farming practices and overexploitation of fuelwood. The most familiar and publicized hazards are those that take place without warning, often of catastrophic dimensions. Human interference has increased the severity of flood in some areas, due to changes in land use such as urbanization and deforestation. In many cases, they are comparable in their potential for damage to pollution, which is a phenomenon created by humans. One such hazard, which has only recently been recognized as widespread, is radon, a radioactive gas associated with certain rock types that seep up through the earth's surface into buildings. The consequences of both natural and human-induced hazards are worst for the earth's creatures and landscape.

Types of Environmental Hazards

Those events or accidents, whether caused by natural processes or human factors, are called environmental hazards, which occur very rarely and aggravate natural environmental processes to cause disaster for human society such as sudden tectonic movements leading to earthquake and volcanic eruption, continued dry conditions leading to prolonged draughts, floods, atmospheric disturbances, collision of celestial bodies etc. Environmental hazards are broadly categorized into two types, namely, natural and man-made. Natural hazards are those which arise from geological or weather-related occurrences whereas man-made hazards, as the term suggests, are caused by man's irresponsible attitude towards nature. This broad classification is based on causative factors.

Figure 9.1 Types of Environmental Hazards

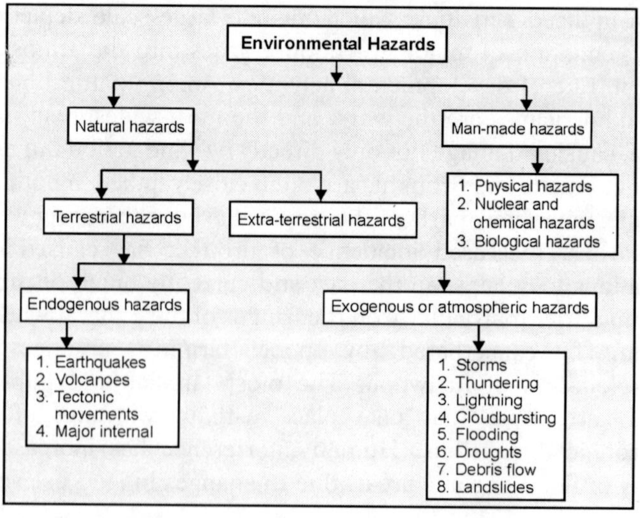

Human-induced activities have tremendous role in degrading environment, particularly in this age of information and technology. Man-induced hazards may be further divided into three sub-categories: physical hazards (forest depletion, landslides, and accelerated soil erosion), chemical and nuclear hazards (release of toxic chemical elements in the air, water and soil, sudden outburst of lethal poisonous gases from chemical factories, nuclear explosions and leakages of radioactive elements from nuclear reactor plants), and biological hazards (sudden increase or decrease of population of species in a given habitat either due to increased nutrients or increase of toxic chemical elements). Similarly, natural hazards are further divided into two broad categories, e.g., (i) terrestrial hazards, and (ii) extra-terrestrial hazards. Terrestrial hazards are categorized into two sub-types, viz., endogenous hazards and atmospheric or exogenous hazards.

Natural Hazards

As stated above, natural hazards occur due to natural phenomena, which vary from very high density of disasters to moderate and low. The high-density hazards are earthquakes, volcanoes and storms, which always have worst effect on the human society. Among the natural hazards, while the terrestrial hazards cause high degree of loss and change the face of the earth's surface, the extra-terrestrial hazards are caused due to collision in celestial bodies and falling of their remains on the surface of the earth. In the history of the earth, there are many instances, when the heavenly bodies seen falling on the earth's surface and damaged the existing landscape. Similarly, the atmospheric hazards are of great intensity ranging from cloud-burst to thundering and lightning. Terrestrial hazards occur due to the two forces of the earth, i.e., endogenous and exogenous forces. Occurrences of earthquakes, volcanoes, and movement of plates and waves inside the earth are called endogenous forces. Exogenous forces include weathering, denudation, and agents of erosion such as rivers, glaciers, winds, oceanic waves, etc. Natural environmental hazards involve comparatively rare high-intensity processes and are caused by both terrestrial and atmospheric processes. The study of natural hazards includes the consideration of identification of specific events, finding of their causative factors, assessment of their impacts on human and other biological communities, prediction of such events and finding their remedial measures.

Terrestrial Hazards

Terrestrial hazards normally include those extreme tectonic events, which are caused by endogenous forces coming from within the earth. In fact, the causative factors of terrestrial hazards are hidden deep within the earth and are not observable by man. Only human beings experience their effects, rather badly. Most of the terrestrial hazards are caused by tectonic movements of lithospheric and oceanic plates relative to each other caused by thermal conditions of the interior of the earth. Among these, volcanic eruptions and earthquakes become more hazardous, which occur after long period of dormant phase.

(a) Endogenous Hazards

Endogenous hazards are those which occur due to the internal forces that come from the earth's interior. These forces are originated due to the presence of high pressure and high temperature, beneath the earth's surface. Movement of the plates as conversion and diversion is also a major cause for origin of these forces. These are more hazardous and disastrous. Although their frequency is less, yet the intensity and dimension of their occurrence is very high, which leads to deadly situation in the affected areas. Endogenous hazards are described as follows:

(i) *Earthquakes:* An earthquake is a major demonstration of the power of the tectonic forces caused by endogenous thermal conditions of the interior of the earth. It is a motion of the ground surface, ranging from a faint tremor to a wild motion capable of shaking buildings apart and causing gaping fissures to open in the ground. The earthquake is a form of energy of wave motion transmitted through the surface layer of the earth in widening circles from a point of sudden energy releases the 'focus'. The Richter scale devised by Charles F. Richter in 1935 measures the magnitude or intensity of energy released by an earthquake. The number indicating magnitude or intensity (M) on Richter scale ranges between 0 and 9 but in fact the scale has no upper limit of number because it is a logarithmic scale. Another scale of the measurement of the degree of destructiveness or intensity of earthquakes is Mercalli scale. The degree of destructiveness or intensity of an earthquake depends on a variety of factors, e.g., magnitude, distance from epicentre, acceleration, duration, amplitude of waves, type of ground, water table, nature of geomaterials of the region concerned and the nature and type of constructions (such as buildings made of wood or bricks or stones or concretes, dams whether concrete or earthen, building made of mud, tin shades, huts, etc.) affected by an earthquake. The place of the origin of an earthquake is called 'focus', which is always hidden inside the earth but depth of which varies from place to place. The deepest earthquake may have its focus at a depth of even 700 km. below the ground surface but some of the major Himalayan earthquakes have their focus around 20–30 km. deep. The place on the ground surface, which is perpendicular to the buried 'focus' or 'hypocentre' recording the seismic waves for the first time is called, 'epicentre'. The seismic waves move away from the source of the earthquake 'focus or hypocentre', in the form of (i) Primary or Pressure waves (P waves), (ii) Secondary, Shear or Transverse waves (S waves), and (iii) Long or Surface waves (S waves). These seismic waves are recorded with the help of an instrument called 'seismography' or 'seismometer' at the epicentre.

(ii) *Volcanoes:* Volcanic eruption is an example of terrestrial hazards but unlike earthquakes, volcanoes are hazards as well as boons to human beings. On the one hand, they kill people and animals and destroy human settlements, agricultural farms, properties through explosive central eruption and spread of hot and liquid lavas coming out of fissure flows and, on the other, they provide rich soils for agricultural purposes. More or less, there is well-marked distributional pattern of volcanoes over the globe. Explosive type of volcanoes wherein volcanic eruption occurs through a central pipe by breaking and blowing of crustal surface due to violent and explosive gases accumulated deep within the earth. Such volcanoes are very destructive. Explosive volcanoes are further divided into five sub-types: (i) Hawaiian type of volcanoes, (ii) Strombolian type of volcanoes, (iii) Vulcanian type of volcanoes, (iv) Pelean type of volcanoes, and (v) Visuvius type of volcanoes. Fissure eruption type of volcanoes are those, which occur in a long fracture, fault and fissure, there is slow upwelling of magma from below, and the resultant lavas spread over the ground surface.

According to the nature of volcanoes, they are categorized into three types, such as active volcanoes, passive volcanoes, and dormant volcanoes. The active volcanoes are those, which occur regularly. They are violent in nature. Passive volcanoes occur in due course of time, while dormant, once occurred, will not occur in the future. The distribution of volcanoes is found in the areas of conversion of plates, where earthquakes have also occurred.

(iii) *Tectonic movements:* Tectonic plates are made of either oceanic or continental crust or the very top part of the mantle, a layer of rock inside the earth. This crust and upper mantle form what is called the lithosphere. Under the lithosphere lies a fluid rock layer called the asthenosphere. The rocks in the asthenosphere move in a fluid manner because of the high temperatures and pressures found there. Tectonic plates are able to float upon the fluid asthenosphere because they are made of rigid lithosphere. Geologists study how tectonic plate's move relative to a fixed spot in the earth's mantle and how they move relative to each other. The first type of motion is called absolute motion, and it can lead to strings of volcanoes. The second kind of motion, called relative motion, leads to different types of boundaries between plates: plates moving apart from one another form a divergent boundary, plates moving towards one another form a convergent boundary, and plates that slide along one another form a transform plate boundary. In rare instances, three plates may meet in one place, forming a triple junction. Current plate movement is making the Pacific Ocean smaller, the Atlantic Ocean larger, and the Himalayan mountains taller.

Divergent plate boundaries occur where two plates are moving apart from each other. When plates break apart, the lithosphere thins and ruptures to form a divergent plate boundary. In the oceanic crust, this process is called seafloor spreading, because the splitting plates are spreading apart from each other. On land, divergent plate boundaries create rift valleys – deep valley depressions formed as the land slowly splits apart. Convergent plate boundaries occur where plates are consumed, or recycled back into the earth's mantle. There are three types of convergent plate boundaries: between two oceanic plates, between an oceanic plate and a continental plate, and between two continental plates. Subduction zones are convergent regions where oceanic crust is thrust below either oceanic crust or continental crust. Many earthquakes occur at subduction zones, and volcanic ridges and oceanic trenches form in these areas. The movement of plates particularly conversion of plates creates severe disturbances in the interior of the earth as well as on the earth's surface as the forms of the earthquakes and volcanoes.

(iv) *Major internal forces:* In the interior of the earth, there are many major forces that originate due to high temperature and high pressure, and change the landscape of the upper crust and are sometimes very much hazardous. One of them is Coriolis force, which is also known as Coriolis acceleration, or an additional force or acceleration acting on the motion of bodies in a rotating system of reference. For example, an object moving above earth in a generally northerly or southerly direction, and with a constant velocity relative to space, will be deflected in relation to the rotation of earth. This deflection is clockwise in the northern hemisphere and counterclockwise in the southern hemisphere. The effect is named after the French physicist Gaspard de Coriolis, who first analysed the phenomenon mathematically. Coriolis forces are of considerable importance in determining prevailing winds and ocean currents and also in the analysis of the flight paths of missiles and rockets.

(b) Atmospheric or Exogenous Hazards

As the name suggests, atmospheric or exogenous hazards are those which are caused due to atmospheric phenomena or changes in it. These hazards are storms, thundering or lightning cloudbursts, flash floods, and consequently mass movement, landslide, debris flow, soil erosion, etc. Unlike the endogenous hazards, atmospheric hazards are also caused due to the natural phenomena and no one can control them. But, some of them, such as landslides and soil erosion, are also caused by the human-induced activities in the form of deforestation, construction of roads, mining, and cutting of terraces for agriculture. These hazards range from heavy losses of property to loss of lives (human and animal). No part of the world has remained untouched by these hazards. These hazards are discussed below:

(i) *Thundering and lightning:* Thundering and lightning are caused due to the collision in two clouds. In other words, when two heavy clouds collide, a peculiar type of sound and light is produced with very high speed. When thundering and lightening reach on the surface of the earth, heavy losses of lives and property take place.

(ii) *Tropical and local storms:* Tropical cyclone, representing closed low-pressure systems, generally having a diameter of about 650 kilometres, is one of the most powerful, destructive, dangerous and deadly atmospheric storms on the planet earth. These are called hurricanes in the North Atlantic Ocean (mainly in the Caribbean Sea and south eastern USA), typhoons in the North Pacific Ocean (mainly in the China Sea, eastern and southern coasts of China, Japan, Philippines and Southeast Asia), cyclones in Bangladesh and eastern coastal areas of India, and willy willy in Australia.

Tropical cyclones become more disastrous natural hazards because of their high wind speed of 180 to 400 kilometres per hour, high tidal surges, and high rainfall intensity (highest recorded rainfall value exceeded 2,000 mm. per day in Philippines). Very low atmospheric pressures causing unusual rise in the sea level and their persistence for several days or say about one week. The total cumulative effects of the high velocities of wind, torrential rainfall and transgression of sea water on to the coastal land become so enormous that the cyclones cause havoc in the affected areas and thus, tremendous loss of human lives and property is the ultimate result of such atmospheric deluge. The 'storm surge' or 'tidal surge' refers to unusual rise in sea level caused by very low atmospheric pressure and the stress of the strong gusty winds on the sea surface.

Cyclonic hazards very often visit the coastal areas of India and the southern coastal areas of Bangladesh. Hurricanes strike the southern and the southeastern coasts of the USA. Gulf coasts of Louisiana, Texas, Alabama and Florida are the worst affected areas. The several local storms of hazardous nature include tornadoes, which affect the southern, and the eastern USA, and thunderstorms. Tornadoes, though smallest in area of all the hazardous atmospheric storms, yet are very horrible to human life and property.

Destruction by tropical cyclones and local storms include loss of human lives and property in terms of destruction of buildings, transport systems, water and power supply systems, disruption of communication system, destruction of agricultural crops, domestic and wild animals, natural vegetation, private and public institutions and so on. Tropical cyclones and tornadoes have a profound impact on human beings. There are six major regions in the world, which are responsible for the origin of the tropical cyclones, viz., West Indies, Gulf of Mexico and Caribbean Sea, Western Pacific Ocean including Philippines islands, China Sea and Japanese islands, Arabian Sea and Bay of Bengal, Eastern Pacific coastal region of Mexico and Central America, Indian Ocean, off Madagascar, and southwest Pacific Ocean, in the region of Samoa and Fiji islands and the eastnorth coast of Australia.

(iii) *Flood and drought hazards:* Also called as cumulative atmospheric hazards, these are caused due to cumulative effects of weather events which prolong for longer period of time ranging from a few weeks to several years depending upon the nature of weather events. For example, any hot day may not become disastrous or hazardous but even very hot and dry days prevail for a few weeks in continuation, environmental hazard in the form of heat waves occurs which affect the environment and human lives, plants and animals. The persistence of exceptionally cold conditions for several days causes severe snowfall.

(a) *Floods:* Flood simply means inundation of extensive land area with water for several days in continuation. Generally, floods are considered to be associated with rivers and people conceive floods as the outcome of accumulation of huge volume of water coming out of the rivers through over topping of riverbanks during peak discharge period. In fact, flood is an attribute of physical environment and thus is a component of hydrological cycle of a drainage basin. It may be pointed out that flood is a natural phenomenon and is a response to rainfall but it becomes hazard when it causes colossal loss of human lives and property.

Floods of the rivers are the responses of both natural and anthropogenic factors. The causes of floods of the alluvial rivers become highly complex and their relative importance varies from place to place. Among the natural factors, which cause river floods, important are prolonged high intensity rainfall, meandering courses of the rivers; extensive flood plains; break in slope in the long profiles of the rivers, i.e., sudden change in channel gradient at the intervening zones of foothill slope of the mountains and upper end of the plains; blocking of free flow of the rivers because of enormous debris provided by landslides and due to volcanic eruptions, nature of river valleys and channel, etc. Anthropogenic activities such as building activity and eventual urbanization, channel manipulation through diversification of river course, construction of bridges, barrages and reservoirs, agricultural practices, deforestation, land use changes, etc., by man invite several hazards in the river system.

(b) *Droughts:* Droughts are more deadly natural environmental hazards because these are directly related to one of the three basic requirements of any form of life, such as air, water and food. Droughts resulting from accumulative effects of water scarcity cause extensive and enormous damage to agriculture and natural vegetation and therefore cause famine and starvation of both human and animal populations of the regions concerned. The meaning and definition of droughts are difficult propositions because there is much variation in the viewpoints and perception of droughts from one region to another and from one group of people to another. Most people are reasonably well aware when a drought situation exists, but it is very difficult to find an overall acceptable definition of drought. It clearly involves a shortage of water, but can really be defined only in terms of particular need. The most common view of drought is of rainfall deficiency, but the links between rainfall and the water, which becomes available to meet demand, is complex. Therefore, definition of a drought relates not only to water needs but also to the complex set of factors involved to supply that need through the hydrological cycle.

Droughts affect all types of life forms in the biospheric ecosystem because both plants and animals directly depend on water. Any shortage of water supply adversely affects them. Thus, the impact of droughts includes ecological, economic, demographic and political aspects. Prolonged drought conditions in a given region change the biotic component of the natural ecosystems because:

- Some species of plants and animals perish, as they cannot withstand extreme drought conditions.
- Some animals migrate to other places and hence, there is marked decrease in the population of certain animal species.
- Some animals die of hunger and starvation.

There is a stiff competition for food due to scarcity created by drought among the animals, which results in the elimination of weaker animals, etc. The most significant ecological impact of prolonged drought is natural control of plant and animal population.

(iv) *Cloudburst and associated events:* In the mountainous regions, cloudburst is a very common phenomenon. When a heavy amount of rainfall occurs in a particular place for a long time (more than one hour), it is called cloudburst. Normally, it occurs in the highly elevated part of the land (mountainous regions). It has deadly consequences. After cloudburst, landslide, debris flow and mass movement are very common in the affected areas. In India, the Himalayan region is the worst prone to cloudburst and its associated events.

Extra-Terrestrial Hazards

Extra-terrestrial hazards occur due to the collision in the celestial bodies. In the solar system, there are nine planets, their satellites, the sun, comets, asteroids and meteorites. They all are rotating in their axis and are making revolution around each other. Many times, they collided with each other as the origin of the earth took place due to collision. These celestial bodies have been falling on the earth's surface causing damage of geomaterials.

Man-made Hazards

As stated earlier, human-induced activities have tremendous role in degrading environment, particularly in this age of information and technology. Man-made hazards may be divided into three sub-categories: physical hazards (forest depletion, landslides, and accelerated soil erosion), chemical and nuclear hazards (release of toxic chemical elements in the air, water and soil, sudden outburst of lethal poisonous gases from chemical factories, nuclear explosions and leakages of radioactive elements from nuclear reactor plants), and biological hazards (sudden increase or decrease of population of species in a given habitat either due to increased nutrients or increase of toxic chemical elements). Man-made hazards are caused due to the following factors:

- Due to population explosion and dependency on the forest for fuel and fodder needs, large-scale forest depletion can be noticed in most of the areas in the world. Forest depletion has severe consequences in the forms of floods, droughts, soil erosion and landslides.
- Increase in greenhouse gases, high level of pollution, and consequently lowering the environmental qualities are the results of human-induced development activities. Everyday, thousand of tons of carbon wastes coming from factories, vehicles, and other means, is causing high level of pollution in the atmosphere.
- Due to high concentration of greenhouse gases in the atmosphere, animal and plant species on the earth's surface are on the brink of extinction everywhere. Using chemical fertilizers in the agricultural land for high production of crops is also responsible for bad human health and several other diseases.

Natural Disaster Reduction and Management

Today, there is growing concern all over the world about the devastation caused by natural disasters which stem from ecological destructive practices and putting humankind in a harmful situation. Many ecosystems have been frayed to the point where they are no longer resilient and able to withstand natural disturbances, setting the stage for 'unnatural disasters' – those made more frequent or more severe due to human actions. By degrading forests, engineering rivers, filling in wetlands, and destabilizing the climate, we are unraveling the strands of a complex ecological safety net (Abramovitz, 2001). Although, the inherent links between disaster reduction and environmental management are recognized, little research and policy work has been undertaken on the subject. The intriguing concept of using environmental tools for disaster reduction has not yet been widely applied by many practitioners (ISDR, 2002).

With an estimated 26 million people in Southeast Asia living below the poverty line, environmental degradation continues to pose formidable challenges of poverty reduction and the achievement of the Millennium Development Goals (MDGs) throughout Asia. The region is home to about half of the world's terrestrial and aquatic resources, but alarming rates of deforestation, degradation of reefs and coastal ecosystems, atmospheric pollution and depletion of freshwater resources continue to lock vast population in downward spiralling cycles of poverty in which the poor pursue unsustainable resource management practices in increasingly fragile environments.

The poverty cycle is exacerbated by frequent natural disasters. In all, 80 per cent of the natural disasters worldwide occur in Asia: and of these, 80 per cent are hydro-meteorological or climate-related. In the decade from 1991 to 2001, natural disasters affected over 1.7 million Asians, costing 369 billion dollars in damage to hard-won assets. It is widely accepted that the poor are the most vulnerable; they suffer the highest number of casualties and have the least capacity to recover. Moreover, costly emergency responses divert limited funds from important environmental management and poverty alleviation initiatives. Meeting the challenges of both environmental degradation and disaster risk remains high on the regional agenda and, more recently, researchers, planners and policy makers have come to recognize how intricately these two factors are related. The International Decade of Natural Disaster Reduction (IDNDR) concluded that "environmental protection, as a component of sustainable development and consistent with poverty alleviation, is imperative in the prevention and mitigation of natural disasters" (ISDR, 2002). Similarly, in 2002, a group of experts from the global change and disaster management communities met in Berlin and prepared a declaration that was presented at the World Summit on Sustainable Development in Johannesburg, South Africa. The declaration pointed to increasing evidence that global environmental change and natural disasters are linked. In 2003, colleagues at the Swedish Embassy in Bangkok sought documentation of this evidence in order to better understand the nature of the linkages and to identify potential synergies between environmental programmes and efforts to reduce disaster risk. More specifically, they ask:

- How can investments in environmental management and sustainable development also reduce disaster risk?
- Is there a prevention dividend that accrues from wise land use planning and development programmes?
- Can prevention dividends be measured; and, how might the ability to estimate these added values enhance policy and programme planning?

Risk and Risk Reduction

Disaster risk reduction represents an important development in both disaster management and sustainable development paradigms. Whereas, earlier generations of disaster managers focused primary attention on the delivery of emergency relief in the aftermath of catastrophes, today, increasing attention is paid to disaster preparedness. Risk may be reduced through a variety of pathways that either target the physical hazards or the underlying socio-economic conditions that create vulnerability. Understanding the interaction of physical and social processes is fundamental to estimating future risk and the value of prevention. The following equation is commonly used to reflect the components of risk:

Risk = Hazards × Vulnerability/Capacity

Risk: The probability of harmful consequences, or expected loss (of lives, people injured, property, livelihoods, economic activity disrupted or environment damaged) resulting from interactions between natural or human-induced hazards and vulnerable/capable conditions.

Hazard: A potentially damaging physical event, phenomenon or human activity, which may cause the loss of life or injury, property damage, social and economic disruption or environmental degradation. Hazards can include latent conditions that may represent future threats. They can be natural in origin (geological, hydro-meteorological and biological) and/or induced by human processes (environmental degradation and technological hazards). Hazards can be single, sequential or combined in their origin and effects. Each hazard is characterized by its location, intensity and probability.

Vulnerability: A set of conditions and processes resulting from physical, social, economic and environmental factors, which increase the susceptibility of a community to the impact of hazards. Positive factors that increase the ability of people and the society they live in to cope effectively with hazards, and also their resilience, or that otherwise reduce their susceptibility, are considered as capacities.

Capacity: The manner in which people and organizations use existing resources to achieve various beneficial ends during unusual, abnormal and adverse conditions of a disaster event or process. The strengthening of coping capacities usually builds resilience to withstand the effects of natural and other hazards.

Resilience: The capacity of a system, community or society to resist or to change in order that it may obtain an acceptable level in functioning and structure. This is determined by the degree to which the social system is capable of organizing itself, and the ability to increase its capacity for learning and adaptation, including the capacity to recover from a disaster (ISDR, 2002).

National development occupies a pivotal position in the cycle as it establishes the underlying socio-economic conditions that determine resilience or vulnerability. Disaster's impacts and the recovery process also affect national development. Accordingly, development is an integral part of disaster reduction programmes.

Reducing Risk

In some countries and in many communities, particularly in disaster-prone areas, prevention has, for generations, been one of the coping strategies for living with risk. Traditional coping strategies are important assets to disaster management programmes. In rural Vietnam, for example, spring planting is timed to bring crops to harvest before the onset of annual flooding. Historically, some governments have also invested heavily in prevention as demonstrated by elaborate system of dykes in Vietnam, which span over 1,000 kilometres and were erected over the course of thousands of years. While, in theory, some disasters may be prevented entirely (for example, some classes of forest fires, landslides and floods), in most cases disaster managers aim to prevent the negative impacts of the hazard rather than eliminating the hazard. An important terminological distinction is made between prevention and mitigation.

Prevention: Activities to provide outright avoidance of the adverse impact of hazards and related environmental, technological and biological disasters.

Mitigation: Structural and non-structural measures undertaken to limit the adverse impact of natural hazards, environmental degradation and technological hazards. Mitigation presumes that the hazard will occur but aims to reduce losses through a variety of interventions. These interventions are generally characterized as either structural or non-structural. Careful timing of agricultural activities is an example of non-structural mitigation, while the construction of dykes is an example of a structural measure.

Figure 9.2 The Disaster Management Cycle

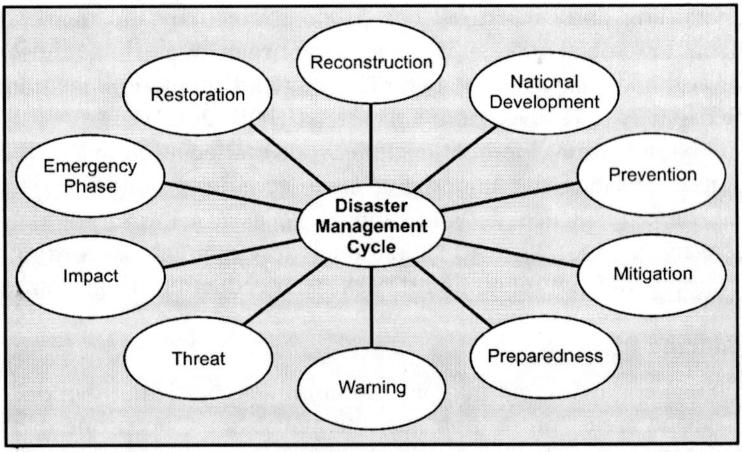

Source: Adopted from Environmental Degradation and Disaster Risk prepared for the Embassy of Sweden/Sida Bangkok by Glenn Dolcemascolo, Asian Disaster Preparedness Centre, February 2004.

Environmental management, which encompasses a broad range of activities, is poised to make significant contributions to disaster risk reduction efforts as well. Advances in both socio-economic and technical approaches to conservation and development have brought some measurable successes in managing forests, wetlands and agricultural lands, water resources and air quality. While some environmental management initiatives aim to control the environment directly, such as reforestation and ecological restoration programmes, most aim to influence the attitudes and behaviours of human communities to engender a culture of sustainability. In many ways, the approach parallels disaster management in acknowledging both the social and the physical dimensions of risk and the critical role of public participation.

Contemporary environmental management is often characterized by a diverse array of managing partners including local villagers, national, provincial and local government agencies, scientists and researchers, NGOs and the international donor community, thus providing for vertical integration. Similarly, horizontal integration of management regimes incorporates the interests and dynamics of multiple sectors such as agriculture, environment, disaster management, tourism and health. Integrated approaches have become widely accepted through the development of new planning paradigms, such as integrated water resources management (IWRM) and integrated coastal zone management (ICZM). Both of these approaches adopt a regional perspective, by taking a watershed, river basin or other ecological region as the basic management unit. Integrated approaches to environmental management are, in principle, well positioned to incorporate

Box 9.1 Ecosystem Functions for Disaster Risk Reduction

Wetlands: Important wetland functions include water storage, storm protection, flood mitigation, shoreline stabilization and erosion control. These functions are also essential for sustainable development.
Forests: Forests play an important role in protecting against landslides, erosion, floods and avalanches. They also protect from droughts.
Coastal Zones: Barrier reefs, barrier islands and mangroves contribute significantly to the mitigation of hurricane risk, storms and tidal surges.

disaster risk reduction in the framework of environmental management and development plans. While the relationship is generally accepted at the conceptual level, more subtle analysis is needed to identify critical thresholds of vulnerability and specific opportunities for synergy.

Endogenous and exogenous events are natural phenomena. These cannot be stopped through human efforts but they can be reduced. The environmental hazards caused due to the human-induced developmental activities can largely be checked with implementation of several management programmes. Though the environmental hazards and disasters, whether natural or man-made, are generally destructive and disastrous to biological communities, both plants and animals in particular, in terms of loss of lives and property, there are a few benefits as well from the natural hazards, e.g., floods renew the soils and enrich their fertility with addition of nutrients; landslides sometimes result in the formation of lakes due to damming of river valleys through debris; volcanic eruptions enrich the soils due to setting down of volcanic ashes; seismic events sometimes cause groundwater barriers which dam the groundwater upslope so that groundwater may be localized. Whatever the positives consequences of these events, they are undoubtedly deadly and devastating and should be managed in time so that the entire humanity, affected by these events, could survive. The reduction of natural hazards and disasters and their management involves:

- provision of immediate relief measures to disaster-affected people,
- prediction of hazards and disasters, and
- measures of adjustment to natural hazards.

Mapping and monitoring of natural hazards and disasters and global changes in environmental conditions are very important aspects of disaster management. This requires in-depth study of hazard-prone areas at global, regional and local levels. The International Council of Scientific Unions (ICSU) and other organizations have launched several research programmes to study the environmental changes caused by human activities and natural disasters in terms of the mechanisms involved in the genesis of such disasters, their monitoring and mitigation.

The United Nations Environmental Programme (UNEP), the Man and the Biosphere Programme (MAB) of the UNESCO, the World Climate Programme (WCP), the International Geosphere Programme (IGP) or Global Change Programme (GCP), the Human Dimension of Global Change (HDGC), the International Social Science Council (ISSC) and the International Federation of Institutes for Advanced Studies (IFIAS), and International Decade for Natural Disaster Reduction are some of the programmes launched for the disaster management.

The following are the main goals of the disaster management programmes:

- To improve the capacity of each country against natural disasters with establishment of early warning systems.
- To foster science and technology aimed at closing critical gaps in the knowledge to reduce natural disasters.
- To disseminate existing and new information related to assessment, prediction, prevention and mitigation of natural disasters.
- To develop measures for the assessment, prediction, prevention and mitigation of natural disasters through various modalities and demonstration projects.
- Eco-friendly developmental programmes should be launched to reduce the man-made disaster equally everywhere in the world, irrespective of the status of the countries, i.e., developed and developing.

Geographical Information Systems (GIS) and aerospace survey may help considerably in the natural disaster reduction and management programmes by providing detailed maps of the problem areas, information about magnitude of the events in an area, planning framework and disaster reduction planning based on past experiences.

CHAPTER 10

Environmental Degradation

Environmental degradation describes a situation in which a part of the natural environment is damaged. It can also be referred to damage to the land, the water or the air. It can also mean a loss of biodiversity or natural resources in an area. Environmental degradation is not new; it has been happening all over the world for centuries but currently the problem has started aggravating. The present level of environmental degradation is so high that it has become a great concern for the entire world community. The greater demands placed on the environment by an ever-increasing human population is putting a great strain and drain on the earth's limited natural resources. Environmental degradation is a serious threat to the lives of people, animals and plants, making it imperative to stop further degradation from occurring.

Types of Environmental Degradation

Most countries are suffering from the menace of various forms of environmental degradation that ranges from the loss of top soil from the earth surface to the high level of pollution. Much influenced by the natural hazards, environmental degradation recently touches higher level due to man-induced activities. Overuse of natural resources, industrialization, urbanization, etc., altogether put a way forward for degradation of environment. Deforestation, desertification, salinization and bush encroachment are the major forms of environmental degradation.

Figure 10.1 Various Forms of Environmental Degradation

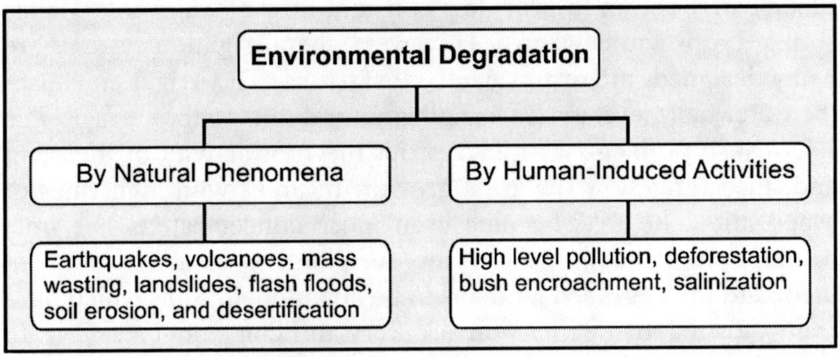

Natural phenomena such as earthquakes, volcanoes, mass wasting, landslides and flash floods, have already been discussed widely in the last chapter. The discussion in this chapter is made on the human-induced activities causing high rate of environmental degradation.

Soil Erosion

Soil erosion is the gradual wearing away of soil by either physical breakdown or chemical solution which is then transported away by means of water, wind or ice to another location. Soil erosion is the leading cause of damage to soils, leaving them barren and ultimately less productive. It can take centuries to create just a few centimetres of soil and only a few moments to destroy the same few centimetres. Today, the rate of erosion has been speeded up by human activities, consequently making soil erosion an ever-increasing problem. Soil erosion results from the ways that people use the land. Practices such as tree felling cause deforestation, and can lead to soil erosion. The removed trees would usually guard the soil from rain and wind as their roots hold the soil in place. Additionally, many land owners cut down trees to create space to plant crops and raise animals which eventually can lead to soil erosion.

Soil Salinization

This is another type of environmental degradation which is particularly common in naturally dry areas that undergo irrigation and do not allow for any fallow periods for the land to recover. Irrigation schemes are set up to provide a constant flow of water to drylands so that crops can be grown. However, when irrigation systems are badly designed, the results can be disastrous. The irrigation causes the water table level to rise bringing natural salts to the surface. The salts cause problems as they restrict the root activity of the plant and, therefore, slow down its growth. In areas with high rates of evaporation the salts become even more concentrated. The final result is that the soils are too salty for plants to be able to grow in them and the degraded land has to be abandoned. Soils which have been affected by salinization are very difficult and expensive to rehabilitate and often remain unused and abandoned.

Desertification

Desertification occurs when productive lands are turned into non-productive desert as a result of poor land management. This generally occurs in semi-arid areas such as the Thar desert of India. Desertification is though a natural process but it also takes place due to the human-induced activities. In India, the Thar desert is spreading towards east of Rajasthan state. Due to global warming and climate change, desertification is spreading speedily worldwide causing severe environmental degradation.

Deforestation

This is the permanent destruction of indigenous forests and woodlands which results in a loss of natural resources as well as a protective barrier for top soil. Deforestation is currently a global phenomenon causing acute environmental problems. In India, at the time of independence in 1947, total forest cover was about 22 per cent which has reduced and presently it is about 19.7 per cent only. This is the case with almost all countries. Equatorial rainforest, which are known for rich biodiversity, are tremendously depleting in Amazon and Kongo basins respectively in Brazil and Central Africa. Similarly, the temperate forest (coniferous) of the northern America and northern Asia that are economically viable are also depleting at an alarming rate.

Bush Encroachment

Bush encroachment happens where woody vegetation gets so thick that it threatens farming lands. Another reason for this is that woody vegetation and grasses have different growth rates leading the woody vegetation to take over and dominate a piece of land. Before the introduction of domestic livestock, the balance between grasses and woody vegetation would have been kept in check by fires and game. This would have resulted in an African savannah dominated by grass with only a few scattered trees. With the introduction of livestock the balance was upset. Most of the game was eliminated and selective grazers were brought in. Fire outbreaks have also been eliminated as far as possible due to human intervention. This means that grasses are heavily eaten but the trees, which are usually controlled by fires, continue to grow. The result is a shift in the balance in favour of trees and woody vegetation. Since the growth of grass is limited, the soil is largely left bare making it especially susceptible to soil erosion by wind and water. The deposits of nutrients are therefore increasingly found only under trees and bushes, making it difficult for grasses to grow. Eventually, the grasses cease to compete for water and die out.

Loss of Biodiversity

Loss of biodiversity is a reduction in the variety of plant and animal species. In areas where environmental degradation has occurred there is often a loss of biodiversity as a result of the disruption to the ecosystem. However, the loss of biodiversity itself can be considered a form of environmental degradation. The range of genetic make-up (plant and animal varieties) in a particular area can be considered to be a natural resource and is important in maintaining a healthy environment. The biodiversity of an area can decrease as a result of pollution, poaching, expanding agriculture and urbanization. Sometimes there is a direct reduction in the number of a particular species which itself, if being threatened, but more often it is as a result of a disruption in the ecosystem and food chain, which causes a domino effect, affecting a greater number of organisms.

Causes of Environmental Degradation

Inappropriate land use can lead to soil degradation. Further, it may take place due to bad farming techniques. Leaving fields bare or ploughing those up and down the sides of a hill can cause severe soil erosion when it rains heavily as the soil has nothing keeping it in place. When the left over parts of crops and animal manure are ploughed back into the soil they serve to replenish and fertilize it. However, if the crops are cut to be fed to animals and the manure is burnt as a fuel, the soil will have no way of replenishing itself, and decreases in fertility. Sometimes landowners make changes in the way they use the land in an attempt to make the land more productive, but often these changes damage the land and actually make it less productive. Overcultivation happens when a farmer does not allow a piece of land to recover in between plantings, exhausting the soil. Left unchecked this can eventually lead to land degradation as the land is being used in a way which is unsustainable. Overgrazing is when more animals than a piece of land can support are allowed to graze in that area. This can cause serious damage to the land. When too many animals are allowed to graze on a piece of land they eat the plants that hold the soil in place. Too many animals may mean that the grass is eaten down to the roots faster than it can grow back. This in turn leads to overgrazing. Overgrazing can cause and accelerate soil erosion and with it there is a loss of soil fertility. When large herds are concentrated around one particular area the animals compact the soil by trampling on it resulting in the soil being unable to retain as much moisture as it needs. Once the animals have overgrazed an area it is often left barren with no protection and the wind blows away the topsoil.

Pollution is also an important factor in causing environmental degradation. Soil can be damaged as a result of waste products and pollutants being deposited and left in it. When garbage from factories, mines and households is dumped in the natural environment, it pollutes the land and leaves its toxins within the soil. The soil therefore becomes unfit to support any plant growth or animal life. The increase in the global population has caused a massive increase in levels of waste and pollution, adding to increasing environmental degradation. It is often the people living in degraded environments are responsible for the damage that has occurred, but this is often as a result of underlying social and economic problems.

Poverty is a major contributing factor to land degradation as it forces millions of people to destroy the resources within their surroundings in order to survive. Poor people often do not have access to the best land, leaving them to depend on the most fragile areas and resources. This means that they have no other choice than to use what resources are available to them, even if these result in the degradation of the land. Foreign debt often forces governments in poor countries to pursue policies and practices which are harmful to the environment in order to keep up with their debt payments, such as intensive farming for export. These are important for the national economies of these countries but may take the place of traditional land uses which may have been more eco-friendly and provide food for people to eat. Besides population growth, natural disasters such as floods and droughts and national emergencies like war and political tensions can also add to the pressures which are placed on the land.

Solutions to Environmental Degradation

Bio-physical

Planting trees over degraded land or afforestation can help to protect the soil from strong wind and from being washed away by soil erosion. If trees are planted in rows along the edges of field, they can be very effective in sheltering both crops and soil which does not have crops in it. Soil salinization and waterlogging can both be avoided by using drip irrigation which delivers only as much water as is needed to a specific area as opposed to high-pressure sprinklers which cause the soil to be permanently covered with water. There are certain traditional farming methods which are better for the land and are more sustainable than some of the methods used by commercial farmers. Rather than monoculture (a practice which involves only the growing of just one crop type), which removes all of the goodness from the soil and leaves crops very susceptible to disease, mixed crops or crop rotation can be a good way to replenishing valuable nutrients in the soil along with maintaining a high level of biodiversity in the area. Effects of overgrazing can be minimized by carefully stocking the land so as to ensure that it is not being grazed by more animals than it can support. It is therefore more likely to stay healthy and to be able to sustain cattle for a longer time. Bush encroachment can be hindered through many processes:

1. Bulldozing is one possible way of stopping bush encroachment, but it is very expensive and labour-intensive. It is also not always as effective as the manual felling of bushes since the trees soon re-grow.
2. Using fires is another method, but this too is also time consuming and labour-intensive and not always effective at killing the trees.
3. Using chemicals is a very risky method of de-bushing as it is non-selective and kills lots of vegetation that helps to support the ecosystem. It is also very costly and is likely to have long-term ecological implications.

Social and Economic

The mass effects of poverty can only be eliminated through the consensus between the world's leading politicians. Without their commitment to end debt, make trade rules fairer, and give more aid to those countries in need poverty cannot be eliminated. Politicians also need to sincerely address the pressing environmental problems that our societies face today, and again commit to policies that enforce the sustainable use of our planet.

An estimated 40 per cent of world deaths can now be attributed to various environmental factors, especially organic and chemical pollutants, according to a study led by David Pimentel, Professor of Ecology and Agricultural Sciences at Cornell. "More and more of us are living in crowded urban ecosystems that are ideal for the resurgence of old diseases and the development of new diseases," said Pimentel, lead author of the *Bio-Science* report titled "Ecology of Increasing Disease: Population Growth and Environmental Degradation." "We humans are further stressed – and disease prevalence is worsened – by widespread malnutrition and the unprecedented increase in air, water and soil pollutants," he further said.

Global climate change will make matters even more worse for humans, the Cornell study predicts. Increased heat favours most human diseases, and also diseases in food crops, and the coming century will see masses of weakened "environmental refugees" fleeing their home areas in a desperate search for food, researchers said.

The Centres for Disease Control and Prevention (CDC) studies conducted at Cornell and other universities find that over a span of planet earth has been made less hospitable to human habitation:

1. Each year, air pollutants adversely affect the health of 4 to 5 billion people worldwide. An expanding world population is burning more fossil fuels, emitting more industrial chemicals and driving more automobiles. The number of cars is increasing three times faster than the rate of population growth.
2. The snail-borne disease schistosomiasis causes an estimated 1 million deaths annually and is expanding its range as human activities provide more suitable habitats in contaminated fresh water. Following construction of Egypt's Aswan High Dam and associated irrigation systems in 1968, prevalence of the Schistosoma mansoni organism in humans in the region increased from 5 per cent to 77 per cent.
3. Of the 80,000 pesticides and other chemicals in use today, 10 per cent are recognized as carcinogens. Cancer-related deaths in the US increased from 3,31,000 in 1970 to 5,21,000 in 1992, with an estimated 30,000 deaths attributed to chemical exposure.
4. Smoke from indoor cooking fires that burn fuelwood and dung is estimated to cause the death of 4 million children each year worldwide.
5. Lack of sanitary conditions contributes each year to approximately 2 billion diarrhoea infections and 4 million deaths, mostly among infants and young children in developing countries. In the US, inadequate sanitation accounts for 9,40,000 diarrhoea infections and about 900 deaths each year.
6. Dengue fever, spread by mosquitoes that breed in old tires and other water holding junk in crowded urban environments, infects an additional 30 million to 60 million people each year.

7. Less than 1 per cent of 500 Chinese cities have clean air. Respiratory disease is the leading cause of death in China.
8. In China, where tobacco smoking increased from approximately 360 to nearly 1,800 cigarettes per person per year, males smoke 98 per cent of the cigarettes. However, mortality due to lung cancer is approximately equal in males and females.
9. Although the use of lead in US gasoline declined since 1985, other sources inject about 2 billion kilograms of lead into the atmosphere in this country each year. An estimated 1.7 million children in the United States have unacceptably high levels of lead in their blood.
10. Production of another gasoline component, the carcinogen benzene that causes leukemia even at low dosages, rose from 0.5 billion kilograms in the US in 1950 to current levels of about 7.5 kilograms per year.
11. The global use of agricultural pesticides rose from about 50 million kilograms a year in 1945 to current application rates of approximately 2.5 billion kilograms per year. Most modern pesticides are more than 10 times as toxic to living organisms as those used in the 1950s.
12. Population-control policies should be combined with effective environmental management programmes. Without international cooperative efforts disease prevalence will continue its rapid rise throughout the world and will diminish the quality of life for all humans.

CHAPTER 11

Population Explosion and Its Consequences

The world countries are experiencing extreme growth of population, resulting in severe impediments of environmental degradation, health hazards and overcrowding in the metropolitan cities. This extreme growth in population is termed as "population explosion". It is a worldwide phenomenon and is staggering and unpredictable in its international and domestic repercussions. However, its intensity varies from area to area. It is most marked in the underdeveloped and industrially less advanced countries.

The growth rate of population and its distribution is uneven in the world, varying from continent to continent and country to country. Water bodies cover about 72 per cent of the earth's area and only 28 per cent of the earth's area is landmass. Out of the total landmasses, 20 per cent land is covered by mountains, where concentration of population is very low. Similarly, many areas of the world have cold and hot deserts. The Antarctica continent and Greenland island are covered by perpetual snow. However, the high concentration of population is limited to the regions of South Asia and Southeast Asia, Central Europe and North America. In terms of population size in the world countries, China ranks first, followed by India where more than 2.5 billion of the world population (7 billion in 2011) live.

The growth rate in the population also varies from the Asian countries to the countries of Europe. Many of the countries of Europe continent are experiencing demographic transition. The United Nations (UN), an accepted authority on population levels and trends, estimates that the world population reached 6 billion in 1999, and is increasing annually by more than 77 million persons. The rate of increase, 1.3 per cent per year, has fallen below the peak rate of 2 per cent per year attained by 1970. By the late 2040s, the UN estimates, the growth rate will be about 0.64 per cent annually, at which time more than 50 countries will experience negative growth.

This unprecedented surge in population, combined with rising individual consumption of food, water and natural resources, has begun to strain the earth's capacity to sustain human life. Demands for water are draining supplies from aquifers and other water sources. Demand for fish – a food staple in many areas of the world – is contributing to the depletion of fish stocks around the world. Human activities that cause pollution and encroach on natural habitats are responsible for the greatest extinction of plant and animal species since the dinosaurs disappeared about 65 million years ago. Meanwhile, global warming has begun to change the earth's climate in ways not yet fully understood. In short, the growth in human population and the scale of human activities appear to be redirecting the natural course of our planet.

Human population growth may be seen to be at the root of virtually all of the world's environmental problems. Increasingly large numbers of people are being added to the world every year. As the number of people increases, more pollution is generated, more habitats are destroyed, and more natural resources are used up. Even if new technological advances were able to cut in half the environmental impact that each person had, as soon as the world's population size doubled, the earth would be no better off than before.

Growth of Population

Although it is not estimated exactly that how much population live on the earth at any given time and what exactly the population growth is, yet the earth is expressing extreme population growth.

Regarding accuracy of population and population growth in large part of the globe, demographic data are non-existent, outdated or unreliable. Most of the African and Asian countries do not have any reliable technique for counting human population. Therefore, to estimate accuracy in the population, data is a huge problem for these countries.

Whatever the situation existed, the data estimated that since the end of the 19th century, the growth of the world's population is of a magnitude that is unparalleled in history. The human race is multiplying at a constantly accelerating rate. According to the best available estimates, the phenomenon started around the middle of the 17th century. Until then, the world's population had risen very slowly and during long periods had either been stagnant or was even on the decline. Between 1650 and the end of the 19th century it probably tripled, going from about 500 to 1,500 million. Between 1920 and 1950, in spite of the enormous blood-letting of a number of catastrophic wars, including World War II, it increased from 1,834 to 2,406 million, or by 31 per cent. Meanwhile, the rate of growth has again increased, amounting now to 17 per cent per decade. For the single year of 1957, an increase of world population was approximately 47 million.

Estimates of world population before 1900 are based on fragmentary data, but scholars agree that, for most of human existence, long-run average population growth approached approximately 0.002 per cent per year (20 per million inhabitants). According to UN estimates, the population of the world was about 300 million in the year and it took more than 1,500 years to reach the 500 million mark. Growth was not steady but was marked by oscillations dictated by climate, food supply, disease and war.

Starting in the 17th century, great advances in scientific knowledge, agriculture, industry, medicine and social organization made possible rapid acceleration in population growth. Machines gradually replaced human and animal labour. People slowly acquired the knowledge and means to control diseases. By 1900, the world population had reached 1.65 billion, and by 1960 it stood at 3.04 billion. Demographic profile of the world's continents, based upon the data of 2002, is shown in Table 11.1 and Figure 11.1.

Table 11.1 Demographic Profile of the World's Continents, 2002

Region	*Births per 1,000 population*	*Deaths per 1,000 population*	*Net number of migrants per 1,000 population*	*Rate of natural increase (%)*	*Growth rate (%)*
World	21.20	8.90	0.10	1.22	1.23
Africa	35.90	14.20	−0.30	2.17	2.14
North America	17.20	7.60	1.50	0.97	1.11
South America	19.70	7.70	−0.20	1.20	1.19
Asia	20.70	7.60	−0.30	1.31	1.28
Europe	10.2	11.40	1.10	−0.12	0.00

In the beginning of the 1950s, a new phase of population growth was ushered in when famine and disease could be controlled even in areas that had not yet attained a high degree of literacy or a technologically developed industrial society. This happened as a result of the modest cost of importing the vaccines, antibiotics, insecticides and high-yielding varieties of seeds produced since the 1950s. With improvements in water supplies, sewage-disposal facilities and transportation networks, agricultural yields increased, and deaths from infectious and parasitic diseases greatly declined. Life expectancy at birth in most developing countries increased from

Figure 11.1 Demographic Profile of the World's Continents, 2002

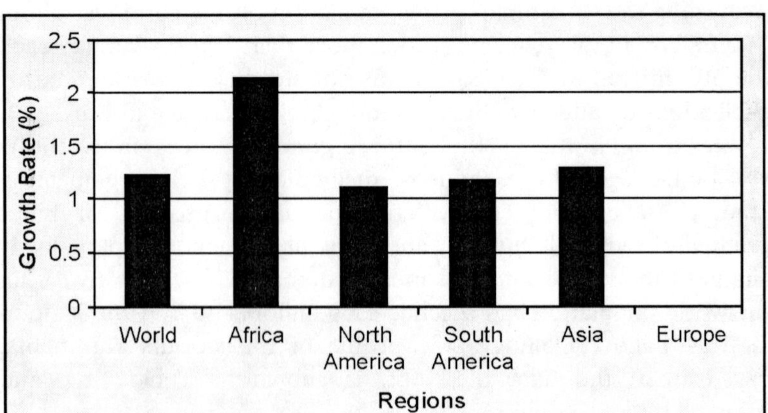

about 35–40 years in 1950 to 66 years by 2000. The rapid decline in deaths among people who maintained generally high fertility rates led to annual population growth that exceeded 3.1 per cent in many developing nations – a rate that doubles population size in 23 years.

In 2000, 1.2 billion people lived in the developed nations of the world, and 4.9 billion people lived in the less-developed countries. Over half the world's population was in East and South Asia: China, with 1.3 billion inhabitants, and India, with some 1.21 (2011) billion were the dominant contributors. Europe and the countries of the former USSR contained 14 per cent, North and South America made up 14 per cent, Africa had 13 per cent, and the Pacific Islands had about 1 per cent of world population.

Differences in regional growth rates are altering these percentages over time. Africa's share of the world population is expected to be more than double by the year 2025. The population of South Asia and Latin America is expected to remain nearly constant; in other regions, including East Asia, the population is expected to decline appreciably. The share of the present developed nations in world population (20% in 2000) is expected to fall to 15 per cent by 2025. Nine out of every ten persons who are now being added to the world's population are living in the less-developed countries. Table 11.2 and Figure 11.2 show the population size and distribution.

The population of the world grew at an alarming rate of about 0.002 per cent per year for most of human history. After 17th century, along with advances in science, agriculture and industry, the world population growth begins to accelerate. Over the next 300 years, the world's population increased five-fold from 500 million in 1650 to about 2.5 billion in 1950. The Population Division of the United Nations predicts that the 6.23 billion humans alive in 2000 will increase to 8.47 billion in 2025 and 10.02 billion in 2050. The UN's estimate assumes that population will peak and stabilize at 11.6 billion in 2200. Others predict that numbers will continue to rise into the foreseeable future – to as many as 19 billion people in 2200.

Table 11.2 Population Size and Distribution

Region	*1750*	*1800*	*1850*	*1900*	*1950*	*2000*	*2050*
Population Size (Millions)							
Africa	106	107	111	133	228	797	1,846
Asia	502	635	809	947	1,437	3,689	5,369
Europe	163	203	276	408	546	727	642
Latin America and the Caribbean	16	24	38	74	166	523	480
North America	2	7	26	82	221	481	722
Oceania	2	2	2	6	12	30	45
Total	791	978	1,262	1,650	2,610	6,247	9,104
% Distribution							
Africa	13.4	10.9	8.8	8.1	8.9	13.1	20.3
Asia	63.5	64.9	64.1	57.4	56.2	60.7	59.0
Europe	20.6	20.8	21.9	24.7	21.4	12.0	7.1
Latin America and the Caribbean	2.0	2.5	3.0	4.5	6.5	8.6	5.3
North America	0.3	0.7	2.1	5.0	8.6	7.9	7.9
Oceania	0.3	0.2	0.2	0.4	0.5	0.5	0.5

Figure 11.2 Population Size and Distribution

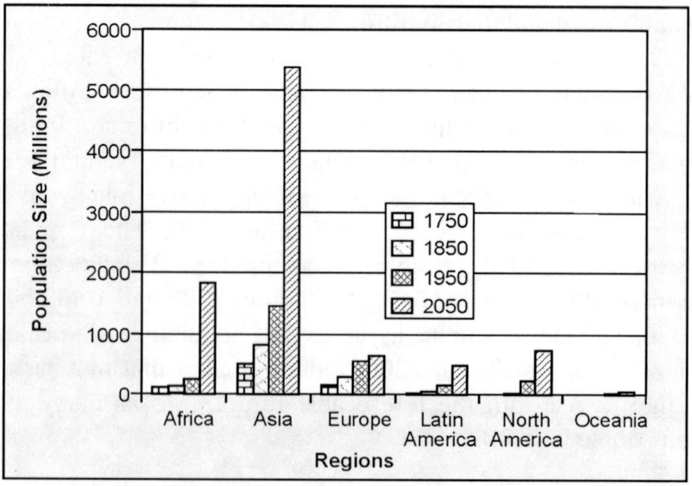

Consequences of Population Explosion

Population explosion is a global phenomenon and its consequences can be seen on the surface and in the atmosphere of the earth. The high growth rate of population, particularly in the developing and underdeveloped countries, led a way for declining environmental quality in all sphere of the earth. Growing population means growing needs. To meet the need of the growing population, there was a tremendous increase in developmental activities. As a result of this, overexploitation of natural resources took place. Overutilization of natural resources, tremendous growth in economic sector and setting up of the large-scale industries worldwide manifested a way for declining biological and natural resources and increasing pollution.

The planet earth is passing through the major environmental problems such as global warming, climate change, ozone depletion, different types of pollution, depletion of natural resource, depletion in fertility of soil, desertification, shrinking forest cover areas and agricultural land, urbanization and acute shortage of water, origin of slums and many more. These problems are worldwide but more intensive in the developing and underdeveloped countries. The origin of all these problems is population explosion. The major consequences of population explosion are as follows:

Overexploitation of Natural Resources

The earth has limited natural resources as a form of water, fauna and flora, minerals, power resources and soils. Out of them, some are renewable and some are non-renewable resources. The non-renewable resources are becoming extinct because of overutilization. Due to the high growth in population, the requirements of the people increased at a tremendous rate and, consequently, the natural resources were unexpectedly overexploited. It is estimated that within 30 years, the reservoir of petroleum will be exhausted completely in the world. Similarly, water scarcity can be observed everywhere. The water crisis and conflict of its sharing within and outside the countries is not a new phenomenon while it took an unprecedented turn and it is a common assumption that the third world war would be fought for water. Along with this, all non-renewable resources are on the verge of vanishing. The scientists worldwide are working hard to find out the substitute of these resources so that the level of requirement can be maintained.

Shrinking Agricultural and Forest Land

There is unbreakable cycle of conversion of forest land into agricultural land and on the same way, the agricultural lands are being converted for the construction of settlements and establishment of industries in or around the cities. The rate of conversion is tremendously high. The tropical rainforests of Amazon basin, within the territory of Brazil, have been cut down at a large scale during the 1980s for construction of agricultural farmlands, which led a way for severe ecological disturbances. In Doon Valley, India, the basmati rice grown agricultural land has been converted into the residential areas at a large scale, which has been resulted in landscape and climate change.

Industrialization and Urbanization

With the advent of industrial revolution in Europe and its further expansion in the other continents, the process of urbanization began. The high growth rate in population in the villages and urban centres further accentuated its magnitude and the cities mostly in the developing and underdeveloped countries are being converted into slums. The urban areas are facing acute shortage of land and water and severe environmental problems. Urbanization is a result of growth of population in urban areas. As a country develops from primarily an agricultural to an industrial economy, large-scale migration of rural residents to towns and cities takes place. During this process, the growth rate of urban areas is typically double the pace of overall population increase. Some 29 per cent of the world population was living in urban areas in 1950; this figure was 55 per cent in 2020. In India, 30 per cent population is living in urban areas (2011). Urbanization eventually leads to a severe decline in the number of people living in the countryside, with negative population growth rates in rural areas. Rapid growth of overall population has deferred this event in most less-developed countries, but it is projected to occur in the early decades of the 21st century.

Environmental Pollution and Enhancement in Greenhouse Gases

Life survives within the better environmental conditions. Air, water, gases and land, which are called life-supporting layers, are very essential for the existence of life. All these given life-supporting layers are utterly polluted. Water is not much useful for drinking and other purposes in the cities because of high pollution. Similarly, air pollution gave birth to many diseases. The upper layer of soil is burnt due to excessive use of chemical fertilizers. It was done for high yield of crops. Population explosion has led to worldwide pollution, which is the major cause of contamination of earth's environment with materials that interfere with human health, the quality of life, or the natural functioning of ecosystems (living organisms and their physical surroundings). Although some environmental pollution is a result of natural causes such as volcanic eruptions, most is caused by human-induced activities. Along with this, an enhancement in the greenhouse gases gave rise to severe problems such as ozone depletion, global warming and climate change. It is noticed that if the existing trend would be continued in the future too, the survival of life on the earth would be in difficult.

Poverty, Malnutrition and Famine

Poverty and malnutrition is a growing and common phenomenon in the underdeveloped countries. High growth rate of population and comparatively low production of crops enhanced the situation of poverty, malnutrition and famine in these countries.

Global Warming

All the environmental problems are caused by the human-induced activities. For instance, the global warming, which is caused by emission of toxic gases and enhancement of greenhouse gases in the atmosphere, is now a global phenomenon. In the route of this phenomenon, the developmental activities, launched by human beings for their welfare, are responsible. Due to global warming, the global average surface temperature has been increasing since 1861, and over the 20th century it has been increased by about 0.6°C.

The Indian Context

The rapid population growth and economic development in the country are degrading the environment through the uncontrolled growth of urbanization and industrialization, expansion and intensification of agriculture, and the destruction of natural habitats. It is against this backdrop that we are today facing the challenge of sustainable development without environmental damage. The existence or the absence of favourable natural resources can facilitate or retard the process of economic development. The three fundamental demographic factors of births, deaths and migration produce changes in population size, composition, distribution and these changes raise a number of important questions of cause and effect. The Population Reference Bureau estimated the world's population in mid-2001 as 6.14 billion. Contribution of India alone to this population was estimated to be 1,033 million. It is estimated that the country's population will increase to 1.26 billion by the year 2016. The projected population indicates that in 2050 India will be the most populous country in the world leaving behind China at second place. The increase in population has been tending towards an alarming situation. India is having 18 per cent of the world's population surviving on 2.4 per cent of its land area and thus has great deal of pressure on its all natural resources. Water shortages, soil exhaustion, deforestation, air and water pollutions afflict many areas. If the world population continues to rise in this indisciplined manner, the impact on environment could be devastating.

The rapid increase of human numbers, coupled with desperate poverty and rising levels of consumption, caused the natural resources to deplete on which the livelihood of present and future generations depend. Poverty is amongst the major consequences of population growth and the people's lifestyle plays a major role in depleting the environment either through its fuel demands for cooking or for earning livelihood for their survival. The growing trends of population and consequent demands for food, energy and housing have considerably altered land use practices and severely degraded India's forests and environment. The growing population put immense pressure on land extensification at the cost of forests and grazing lands because the demand of food could not increase substantially in comparison to population. Thus, horizontal extension of land has fewer scopes and relies mostly on vertical improvement that is supported by technical development in the field of agriculture, i.e., HYV seeds, fertilizers, pesticides, herbicides and agricultural implements. The relationship between population growth, resource depletion and environmental degradation has been a matter of discussions for decades. Both population growth and non-sustainable development are cause for concern in India. Though the relationship is complex, population size and growth tends to expand and accelerate these human impacts on the environment.

Population Growth in India

India is the second most populous country in the world after China. Recently, the population of India has crossed the one billion mark. According to the Census of India 2001, the population of India in 2001 was 1,027 million, while in 2011, the total population is 1,210 million. At the time of independence, the country's population was 342 million. The number has multiplied three-fold in around five decades. The total population size of India has grown from 361 million in 1951 to around 1,210 million in 2011. The population of India increased by three times during the period from 1951 to 2011. The rural population of India has increased around two and half times from 298.7 million to 741.7 million during the same period, whereas the urban population has grown 4.6 fold from 62.4 million in 1951 to 285.3 million in 2001. The decadal growth rates of the population have been irregular – from 13.31 per cent in 1951 to 24.8 per cent in 1971. It declined to 24.7 per cent in 1981, 23.8 per cent in 1991 and 21.35 per cent in 2001. The rural decadal growth rate of population varies from 8.79 per cent in 1951 to 17.97 per cent in 2001, whereas the urban decadal growth rate varies from 41.43 per cent in 1951 to 31.11 per cent in 2001.

There are various reasons for this variation in the trend of population growth rate in various censuses. The increase in population has been due to the improvement in health conditions and control of diseases. The density of population has gone up from 117 in 1951 to 324 in 2001 and it always shows an increasing trend over the census years. Several push and pull factors are presumed to be operative towards distress outmigration from rural to urban areas. This might be due to the declining resource availability per capita and shrinking economic opportunities in rural areas, and better economic opportunities, health and educational facilities, etc., in urban areas, providing opportunities for higher level of human capital development could be the underlying factors for rural outmigration.

Fertility and Mortality Rates

The growth in population depends upon fertility, mortality and migration. The processes of accelerated population growth in India till 1970s were observed to witness a marginal deceleration during 1980s, as the decadal population growth was 23.9 per cent compared to 24.9 per cent in 1970s. Further, declines in fertility have been witnessed during 1990s and 2000s. Birth rate in India has declined from 33.9 per thousand populations in 1981 to 29 per thousand populations in 1991 and it further declined to 25 per thousand populations in 2001. At the same time crude death rate has also declined from 12.5 per thousand populations in 1981 to 9.8 per thousand populations in 1991 and it further declined to 8 per thousand populations in 2001. Thus, the natural growth rate of India's population has declined from 21.4 per cent in 1981 to 17 per cent in 2001. The total fertility rate has also declined from 4.5 children per women in 1981 to 3.1 children per woman in 2001. Similarly, the infant mortality rate has steadily declined from 110 in 1981 to 72 in 1991 per thousand populations but it is still very high (i.e., 66 in 2001).

Trends in Poverty and its Environmental Effects

Most of India's poor live in rural areas and are engaged in agriculture. India's anti-poverty and employment generation programmes, along with overall economic growth planning efforts, have helped reduce the poverty ratio in the country. The people below poverty line declined from 55 per cent in 1973 to 26 per cent in 1999–2000. The absolute number of poor has, however, declined from 320 million in 1973–74 to 260 million in 1999–2000. Over the period from 1987–88 to 1999–2000, urban and rural poverty declined but more declines have been experienced by urban areas.

Poverty is said to be both cause and effect of environment degradation. Poorer people, who cannot meet their subsistence needs through purchase, are forced to use common property resources such as forests for food and fuel, pastures for fodder, and ponds and rivers for water. It also contributes to environmental degradation through overexploitation of natural resources like land, air and water. Population pressure driven overexploitation of the surface and underground water resources by the poor has resulted into contamination and exhaustion of these resources. Urban population also uses rivers to dispose of untreated sewage and industrial effluent. The result is that health of those dependents on untreated water resources is increasing at risk. Moreover degraded environment can accelerate the process of impoverishment, again because the poor depend directly on natural assets. The poverty and rapid population growth are found to co-exist and thus seems to reinforcing each other. Poverty also affects the demographic characteristics of the population and hinders the transition to slower population growth. The deterioration of natural resources and unsafe living conditions affect the environment and health of the poor people.

Environmental Challenges

Population growth and economic development are contributing to many serious environmental problems in India. These include pressure on land, land/soil degradation, depletion of forests, habitat destruction and loss of biodiversity, changing consumption pattern, rising demand for energy, air pollution, global warming and climate change, and water scarcity and water pollution.

Pressure on Land

India faces the most acute pressure on agricultural land. Today, every million hectares of land supports 7.27 million people. Forty three per cent of the land is under cultivation, which is one of the highest in the world. A change in land utilization pattern implies an increase or decrease in the proportion of area under different land uses at a point in two or more time periods. Over the past fifty years, while India's total population increased by about three times, the total area of land under cultivation increased only 20.27 per cent from 118.75 million hectares in 1951 to 142.82 million hectares in 2001. Most of this expansion has taken place at the expense of forest and grazing land. Despite expansion of the area under cultivation, less agricultural land is available to feed each person in India. It shows variations in land use and a narrow range of fluctuations in the proportion of net sown area to total land in the country since 1951. Out of total geographical area of 329 million hectares, only 306 million hectares is the reporting area. The land for non-agricultural uses (housing, industry and others) has increased from 9.36 million hectares in 1951 to 22.97 million hectares in 2001. More than 19.4 million hectares are snowbound and remote leaving only 237 million hectares for agriculture, forestry, pasture and other biomass production. The area under cultivation had increased by about 30 per cent until 1981 and thereafter depicts marginal decline. The net sown area increased from 119 million hectares in 1950–51 to 140 million hectares in 1970–71 mostly through reclamation of old fallow and culturable wastelands and diversion of groves. Later, the net area sown has increased only marginally from 140 million hectares in 1970–71 to 141 million hectares in 2000–01, indicating that the private efforts have peaked and the intervention of the government is required for further land reclamation.

The extent of agricultural intensification is characterized by increase in cropping and irrigation intensity and higher use of chemical fertilizers, pesticides and insecticides. The process of agricultural intensification is leading to land degradation, overexploitation of underground water resources, and increased use of chemical fertilizers leading to eutrophication and water pollution. Agricultural intensification because of increasing cropping intensity, irrigation intensity and excessive use of chemical fertilizers resulting into waterlogging, salinization and alkalinization of croplands and eutrophication of water bodies and ill-health of oceans and thus reductions in biodiversity.

Land Degradation

Direct impacts of agricultural development on the environment arise from farming activities, which contribute to soil erosion, land salination and loss of nutrients. The spread of green revolution has been accompanied by overexploitation of land and water resources and use of fertilizers and pesticides have increased manifolds. Shifting cultivation has also been an important cause of land degradation. Leaching from extensive use of pesticides and fertilizers is an important source of contamination of water bodies. Intensive agriculture and irrigation contribute to land degradation, particularly salination, alkalization and waterlogging. It is evident that most of the land in the country is degrading, thus affecting the productive resource base of the economy. Out of the total geographical area of 328.7 million hectares, 175 million hectares are considered to be degraded areas. Water and wind erosion is the major contributor of 141.3 million hectares to soil erosion, with other factors like waterlogging 8.5 million hectares, alkali soil 3.6 million hectares, acid soil 4.5 million hectares, and saline soil including coastal sandy areas 5.5 million hectares adding to the degradation of land. While soil erosion by rain and river in hill areas causes landslides and floods, deforestation, overgrazing, traditional agricultural practices, illegal mining and setting up of development projects in forest areas have resulted in opening up of these areas to heavy soil erosion. While ravines and gullies reported 4 million hectares, area subject to shifting cultivation reported 4.9 million hectares and riverine and torrents erosion due to floods and eutrophication due to agricultural run off reported 2.7 million hectares. The increasing intensification and extensification also results in salination, alkalization and waterlogging in irrigated areas of the country. For achieving and maintaining food security, sustainable forestry, enhanced agricultural production and controlling of land/soil erosion are very much necessary.

Declining per Capita Forest and Agricultural Land

The population growth has resulted in a downward trend in per capita availability of forest and agricultural land since the 1950s. Per capita availability of forests in India is much lower than the world average. Overall, the per capita availability of forest land had oscillated around 0.113 hectare during the 1950s, and then has consistently declined from 0.124 hectare in 1960–61 to 0.071 hectare in 1998–99 – a level that is extremely low compared to the world standards. The growth of population is expected to be faster than hopes for improvements in forest cover as well as quality. Over the last ten years, despite governmental initiatives of joint forest management, tree grower's co-operative movements and other efforts, tangible results are still to be observed, and forest depletion and degradation is still increasing. Similarly, the per capita availability of agricultural land in rural areas has declined consistently from 0.638 hectare in 1950–51 to 0.271 hectare in 1998–99 and is expected to decline further as population continues to grow.

Habitat Destruction and Loss of Biodiversity

Protection of earth's biological diversity is an important goal. Biodiversity has direct consumptive value in food, agriculture, medicine and industry. It also has the aesthetic and recreational values. The greatest threat to biodiversity is not from the destruction of plants and animals alone, but from the destruction of their habitats. India is one of the 12 mega-biodiversity countries of the world. From about 70 per cent of the total geographical area surveyed so far 46,000 plant species and 81,000 animal species representing about 7 per cent of the world's flora and 6.5 per cent of the world's fauna have been discovered. Population growth leads to expanding human settlements and increasing demand for food, fuel and building materials. Modernization of agriculture also threatens potentially valuable local crops. Biodiversity the world over is on the brink of extinction because the habitats are threatened due to such development programmes as creation of reservoirs, mining, forest clearing, lying of communication and transport networks. It is estimated that in the worldwide perspective slightly over 1,000 animal species and sub-species are threatened with the extinction rate of one per year, while 20,000 flowering plants are thought to be at risk (Compendium of Environment Statistics, 2000).

Increasing Vehicles

The economic and industrial development is inevitably accompanied by changing patterns of consumption. The number of registered motor vehicles in India provides one useful indicator of expanding consumption and economic growth. The increasing vehicles in country, producing more air pollution, fuel consumption, traffic jams and demands for road construction – often at the cost of agricultural land. The total number of registered vehicles in India has increased from 3 million in 1950–51 to 55 million in 2001–02. The number of registered two-wheelers rose from just 0.27 million in 1950–51 to more than 385 million in 2000–01. The number of cars, jeeps and taxis also registered a massive increase from 1.59 million in 1950–51 to 71 million in 2000–01. The number of registered trucks and buses also registered a massive increase from 0.82 million in 1950–51 to 2.95 million in 2000–01 and 0.34 million in 1950–51 to more than 0.63 million in 2000–01, respectively. However, the major share is contributed by metropolitan cities in all registered vehicles in the country.

The population of India in 2000 was just over 1 billion, and there were about 10 motor vehicles for every 1,000 people, or a total of roughly 10 million motor vehicles in the country. In 2020, the population will be above 1.3 billion, and there will be about 44 motor vehicles for every 1,000 people, making a total of 57 million vehicles (Energy Information Administration, 2001). An increase in vehicular pollution is associated with a number of environmental problems like air pollution and global warming. In most urban areas of the country, air pollution has worsened due to traffic congestion, poor housing, poor sanitation and drainage and garbage accumulation. The environmental effects of fuels like oil and petroleum products are of growing concern owing to increasing consumption levels.

Rising Demand for Energy

The environmental effects due to increasing consumption levels of fuels like coal, lignite, oil and nuclear are of growing concern to various researchers. The combustion of these fuels in industries has been a major source of pollution. Coal production through open cast mining and its supply to and consumption in power stations and industrial boilers leads to particulate and gaseous pollution, which can cause pneumoconiosis, bronchitis and respiratory diseases. Both energy production and consumption have increased steadily in India since independence. The production of coal and lignite has increased from 32.2 million tons in 1950–51 to 313.70 million tons in 2000–01 – an increase of 9.74 times. The production of petroleum products registered an increase of 29 times – from 3.3 million tons in 1950–51 to 95.6 million tons in

2000–01. The bulk of commercial energy comes from the burning of fossil fuels, viz., coal and lignite in solid form, petroleum in liquid form and gas in gaseous form. In addition to emission of greenhouse gases, the burning of fossil fuels has not only led to several ecological problems but also health problems like cancer risk, respiratory diseases etc.

Burning of traditional fuel adds a large amount of carbon dioxide into atmosphere and increases air pollution. The production of electricity has increased from 5 billion KWH in 1950–51 to about 500 billion KWH in 2000–01. The shares of thermal power and hydropower have also altered substantially. The share of thermal power has increased from 51 per cent in 1950–51 to about 82 per cent in 2000–01 whereas the share of hydropower declined from 49 per cent in 1950–51 to about 15 per cent in 2000–01. The share of nuclear power is nominal. This clearly indicates that burning of fossil fuels, especially coals, emits lot of carbon dioxide in the atmosphere and leads to global warming. The increasing population numbers and growing affluence have already resulted in rapid growth of energy production and consumption in India, and this trend can only be expected to accelerate in the future. A considerable amount of air pollution results from burning of fossil fuels. Moreover, the resources for fossil fuels are also limited and, thus, the exploration of alternative energy resources would be the only way out.

Policy and Planning

Conservation of natural resources for sustainable development and sustenance of mankind demands strict policy and planning to be framed and implemented. There is also a need to control poverty and population growth below replacement level in the country. Unless significant measures are taken to incorporate environmental concerns into agricultural development, urban planning, technological innovations, industrial growth, and resource management, the situation is not likely to improve in the future. Control on pollution of all types for a healthy living is essential. Special efforts should be made for informing and educating the people and local leaders about the adverse effects of large population through specially designed information, education and communication

(IEC) activities. In order to increase green cover and to preserve the existing forests, afforestation and social forestry programmes should be implemented at local level. It is vital to take preventive and curative measures to control water pollution caused due to chemical fertilizers, pesticides and other wastes. Wastewater treatment plants should be established in accordance with the need of time and its usage should be encouraged. The heavy penalty should be imposed on industries disposing of the wastes into the rivers. Moreover, the landfills are to be properly managed to prevent groundwater contamination. More emphasis should be laid on compulsory environmental education at the school level in order to make people aware of the environment protection. The environment protection should not be the responsibility of government alone but local people and leaders should also be encouraged to make dedicated efforts to eradicate the environmental problems.

Population explosion – the rapid and dramatic rise in the world population has occurred over the last few hundred years. The world's population increased from 1.65 billion in 1900 to 3.02 billion in 1960. Thus, the size of the population nearly quadrupled in the span of 100 years – a historically unprecedented rate of increase. Most of this growth is now taking place in the world's developing countries, where rates of natural increase are much higher than those in industrialized developed countries. Although, it is true that rates of population increase are much slower in the developed world than in the developing world, it would be a mistake to assume that the population growth problem is primarily a problem of developing countries. In fact, because larger amounts of resources per person are used in the developed nations, each citizen from the developed world has a much greater environmental impact than does a citizen from a developing country. Conservation strategies that would not alter lifestyles but would greatly lessen environmental impact are more essential in the developed world.

Evidence now exists suggesting that the most important factors necessary to lower population growth rates in the developing world are democracy and social justice. Studies also show that population growth rates have fallen in areas where several social conditions have been met. In these areas, literacy rates have increased, and women are given economic rights equal to that of men and thus are able to hold jobs and own property; also, birth control information is more widely available, and women are free to make their own reproductive decisions. There is thus crucial need of controlling population growth as many of the countries in the world have already started this process and succeeded. In a similar manner, both the developing and underdeveloped countries have to initiate and implement the way for controlling growing population, otherwise, it will lead to severe environmental problems.

CHAPTER 12

Global Warming and Climate Change

The world countries are confronted with the severe duo impediments of global warming and climate change that are the most crucial atmospheric phenomena observed in the entire world during the recent period. Alongwith the advancement of technology and cultural transformation, the race for economic development was enormous during the last centuries. As a result of this, the concentration of the greenhouse gases increased in the atmosphere. It is well known that the atmosphere is made of various gases, dust particles and water vapours. Among the gases, nitrogen and oxygen, which make about 99 per cent of the total atmospheric gases, are the most abundant and important gases for the sustenance of life on the earth. The other gases are carbon dioxide methane, nitrous oxide, carbon monoxide and chlorofluorocarbon. These gases are called greenhouse gases. Following the invasion of science and technology in all sectors of development, greenhouse gases are increasing at an alarming rate. As a result, not only the temperature is increasing but the percentage of oxygen in the atmosphere is also receding. Owing to burning of fossil fuels like petroleum and coal and also due to forest fires, the carbon dioxide is increasing. Furthermore, following the industrialization, carbon dioxide has increased about 25 per cent in the atmosphere. It is estimated that during the next 50 years, the level of oceanic water would be raised around 2–3 metres and most of the coastal regions would be half submerged. The Antarctica continent, which is covered by snow, is melting and if the process would be continued during the coming period, the rise in oceanic water could be alarming. A survey report by British Antarctica reveals that around 2.5°C temperature increased in the coastal region of Antarctica. The ozone layer – the life-protecting layer – is also being depleted. All this has resulted in severe climate change in terms of extremity in the seasons along with variability from one geographical location to other. The summers are now extremely hot, winters excessively cold, and floods and droughts are also severe.

Definition of Global Warming

Global warming is a process by which the average temperature of the atmosphere, oceans and landmasses of the earth is increasing. It is a burning issue in the present era of globalization and it gives a sense of increasing temperature over the surface of the earth and in the atmosphere. Still, we do not have any standard unit to measure the rate of warming but there are several symptoms, such as climate change, which indicate the rise in temperature. The planet earth has warmed and cooled many times during the 4.65 billion years of its history. At present, it appears to be facing a rapid warming, which most scientists believe is the result, at least in part, of human-induced developmental activities. The basic cause of global warming is thought to be the burning of fossil fuels, such as coal, oil and natural gas, which releases carbon dioxide and other substances into the atmosphere and are known as greenhouse gases. As these gases increase in the atmosphere, it becomes a better insulator, retaining more of the heat provided to the planet by the sun.

As early as 1896, scientists suggested that burning fossil fuels might change the composition of the atmosphere and that an increase in global average temperature might result. The first part of this hypothesis was confirmed in 1957, when researchers working in the global research programme called the International Geophysical Year sampled the atmosphere from the top of the

Hawaiian volcano Mauna Loa. Their instruments indicated that carbon dioxide concentration was indeed rising. Since then, the composition of the atmosphere has been carefully tracked. The data collected show undeniably that the concentrations of greenhouse gases in the atmosphere are increasing. Scientists have long suspected that the global climate, the long-term average pattern of temperature, was also growing warmer, but they were unable to provide conclusive proof. Temperatures vary widely all the time and from place to place. It takes many years of climate observations to establish a trend.

The concentration of greenhouse gases is increasing near the earth's surface and consequently the temperature is rising. But does the increase in gases necessarily cause the warming, and will these two phenomena continue to occur together? In 1988, the United Nations Environment Programme and the World Meteorological Organization established a panel of 200 leading scientists to consider the evidence. The panel concluded that global air temperature had increased 0.6°C since 1861. The panel agreed that the warming was caused primarily by human-induced developmental activities that add greenhouse gases to the atmosphere.

Box 12.1 Salient Features of Global Warming

1. The global mean surface temperature has increased by a substantial margin of 0.6°C since the pre-industrial times.
2. The concentration of greenhouse gases in the atmosphere has been significantly increasing at a rate between 0.25 and 5.0 per cent per annum.
3. The global temperature has risen in past two decades faster than in any other 20 years period.
4. Nighttime temperature overland has generally increased more than daytime temperature.
5. The 1990 to 1999 warm phase of the El Nino/Southern Oscillation (ENSO) was unusual in context of past 120 years.
6. The frequency as well as intensity of extreme weather events has become more variable.

Cont'd...

...Cont'd

7. There has been a marked tropospheric depletion of ozone gases and increase in greenhouse gases near the earth's surface.
8. Recent years have been the warmest since 1860 (period of instrumental records).
9. Magnitude and pattern of climate change and variability are not uniform throughout the world.
10. Temperature is likely to rise by about 2°–4°C in winter and 1°–2°C in summer by 2030. However, there will be large inter-regional variations and some regions may witness cooling trends.
11. The diurnal range of temperature has increased.
12. With increasing temperature and evapotranspiration, soil moisture is likely to reduce by 0.2 cm after 100 years.
13. There is likelihood of shorter duration monsoon with more intensive rainfall.
14. The frequency and intensity of extreme weather events have an increasing trend.
15. The frequency of depressions and cyclones in Indian seas has decreased.
16. Simultaneous increase in temperature and rainfall over India is unlikely.

Causes of Global Warming

The major causes of global warming and consequent climate change are El Nino process (due to it, severe droughts are observed, particularly in the South Asian countries in 1987), burning of fossils fuels (in 1992–93 winters, the carbon contains were found in the snow in the Himalayan region after Iraq's invasion of Kuwait and the burning of oil wells), industrialization, depletion of forests (northern coniferous forests, which are extended in north Canada, Alaska and Siberia, depleting at a large scale due to human interference and similarly, tropical forests of equatorial region, particularly of Brazil, are cut rigorously by the government and these forestlands are being converted into agricultural land, and consequently, the region is facing acute health problems and imbalance in the atmospheric phenomenon), depletion of ozone layer (there is a hole in the ozone layer, about 11 million sq. km. as the area of Europe Continent, between Oceania Continent and Antarctica (the Southern Pole), and urbanization and mounting population (unlimited extension of urban areas resulted in increase in vehicles and origins of slums). We can now discuss these factors in some details in the following paragraphs:

Figure 12.1 Causes of Global Warming

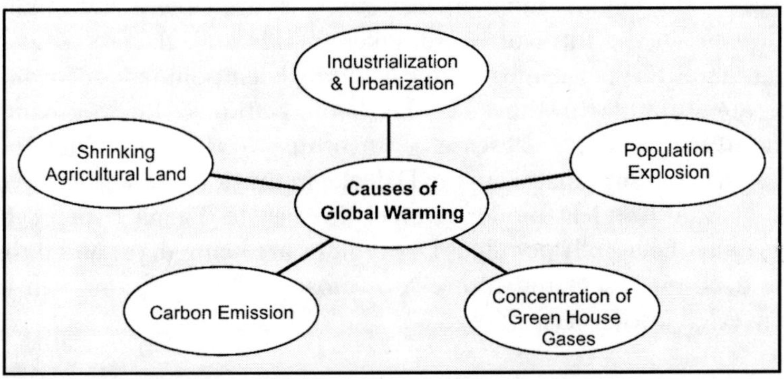

Industrialization

Industrialization is one of the prominent causes for global warming. Due to advances in science and technology, the growth rate of industrialization has gained momentum in both developed and developing countries. The smog releases from the industries is very harmful for the structured atmosphere. Along with industrialization, increasing numbers of vehicles are also releasing toxic smog in the atmosphere and, as a result of this, the increasing rate of carbon dioxide, CFCs and carbon monoxide is very high. Methane, nitrous oxide and surface ozone gas (in and around the cities) are also increasing sharply and causing great harm to life.

Population Explosion

Increasing human population is the prime concern of the environmentalists because due to this factor, environment is deteriorating at a large scale. For fulfilling daily needs and increasing development activities, the human population is heavily dependent on natural resources. Overutilization of natural resources has thus led to the degradation of environment. Pollution, on the other hand, raises a question on the longevity of human race. The metropolitan cities of most of the developing countries are converting into slums and has resulted in the spread of epidemics among the urban population. In India, according to a survey made by the Central Board of Water Pollution Control (CBWPC), 13 cities are fully affected by water pollution. These are all industrial cities. The river water in these cities is neither usable for drinking nor for irrigation. About 190 billion cubic metre water is available for drinking purposes in the form of rivers, lakes, ponds and glaciers. Out of that, about 70 per cent of water has fully been polluted, according to an estimate by the scientists of National Environmental Engineering and Research Institute (NEERI). Planning Commission says that from the Dal lake in the north to Periyar river in the south and Damodar river in the east to Thana river in the west have been fully polluted. Dry regions are being dryer and if this trend continues in the future too, most of these regions will be converted into deserts.

Shrinking Agriculture Land

Both urbanization and increasing population are paving the way to convert agricultural land into settlements. This trend can be seen everywhere, mostly near the towns and cities. Many of the countries of the world, where earlier vast agricultural land appeared has now converted into industrial areas, urban agglomerations and residential areas, etc. To search out an additional agricultural land for livelihood, mass cutting of forest lands occur. As a result of this, the present impediment of global warming and climate change can be seen everywhere.

Depletion of Forestland

Tropical rainforest of the world are depleting due to mass cutting of these forests. A report by the FAO reveals that the tropical rainforests are being cut down at the rate of 15.4 million hectares, because forestland is converting into agricultural land. It is also due to population growth and increasing need for timber for industrial and other development purposes. Presently, the total land cover under forests in India is 20.6 per cent; while in Brazil, it is 57 per cent, in Russia 45 per cent and in America and Canada it is 33 per cent. The per capita forestland in India is 0.2 hectare whereas in Russia, it is 3.5, in America 1.8 and in Sweden it is 4 hectare.

When we see the forest map of India, we find the area under forest cover differs from 2.8 per cent in Delhi to 86.90 per cent in Andman & Nicobar Islands. The following table reveals the per cent of forest cover area in different states of India:

Table 12.1 Forest Cover Area in Different States of India

Forest Area	*States of India*
More than 50%	Andman & Nicobar Islands, Manipur, Nagaland, Mijoram, Tripura, Arunachal Pradesh, Jammu & Kashmir, and Uttaranchal
33–50%	Assam, Himachal Pradesh, Madhya Pradesh, Meghalaya, Orrisa, Sikkim, and Dadar & Nagar Haveli
20–33%	Andhara Pradesh, Kerala, Maharashtra, Goa and Daman Dweep
10–20%	Bihar, Karnataka, Rajasthan, Tamilnadu, Uttar Pradesh and West Bengal
Below than 10%	Gujarat, Haryana, Punjab, Delhi and Chandigarh

Depletion of forest areas has severe implications on climate change and global warming. According to the researchers, the impact of forest depletion also falls on the wildlife. A report depicts that due to the presence of dense forest in an area or region, 10°C temperature recedes in the summers and 10–20 per cent rainfall increases during the rainy season.

Consequences of Global Warming

Scientists use elaborate computer models of temperature, precipitation patterns and atmosphere circulation to study global warming. Based on these models, scientists have made several predictions about how global warming will affect weather, sea levels, coastlines, agriculture, wildlife and human health. According to a study, the following consequences may take place due to global warming:

1. The rise in temperature is expected to melt polar ice caps and glaciers as well as warm the oceans, which will in turn expand ocean volume and raise sea level by 9 to 100 cm., flooding some coastal regions and even entire islands. Some regions in

warmer climates will receive more rainfall than before, but soils will dry out faster between storms. This soil desiccation may damage food crops, disrupting food supplies in some parts of the world. Plant and animal species will shift their ranges towards the poles or to higher elevations seeking cooler temperature and species that cannot do so may become extinct.

2. Scientists predict that during global warming, the northern regions of the Northern Hemisphere will heat up more than other areas of the planet, northern and mountain glaciers will shrink, and less ice will float on northern oceans. Regions that now experience light winter snows may receive no snow at all. In temperate mountains, snowlines will be higher and snow packs will melt earlier. Growing seasons will be longer in some areas. Winter and nighttime temperatures will tend to rise more than summer and daytime ones.
3. The warmed world will be generally more humid as a result of more water evaporating from the oceans. Scientists are not sure whether a more humid atmosphere will encourage or discourage further warming. On the one hand, water vapour is a greenhouse gas, and its increased presence should add to the insulating effect. On the other hand, more vapours in the atmosphere will produce more clouds, which reflect sunlight back into space, which should slow the warming process.
4. Greater humidity will increase rainfall, on average, about 1 per cent for each Fahrenheit degree of warming (rainfall over the continents has already increased by about 1 per cent in the last 100 years). Storms are expected to be more frequent and more intense. However, water will also evaporate more rapidly from soil, causing it to dry out faster between rains. Hurricanes, which gain their force from the evaporation of water, are likely to be more severe. This scenario may also take us to very cold periods to occur.
5. Warmed globe will probably produce as much food as before, but not necessarily in the same places. More insects and plant diseases may also afflict crops and woodlands.
6. Animals and plants will find it difficult to escape from or adjust to the effects of warming because humans occupy so much land. Under global warming, animals will tend to migrate

towards the poles and up mountainsides towards higher elevations, and plants will shift their ranges, seeking new areas, as old habitats grow too warm. In many places, however, human development will prevent this shift and species may disappear.

7. In a warmer world, scientists predict that more people will get sick or die from heat stress, due to hotter days than warmer nights (giving the sufferers less relief). Diseases now found in the tropics, transmitted by mosquitoes and other animal hosts, will widen their range as these animal hosts move into regions formerly too cold for them. Scientists also predict rising incidence of allergies and respiratory diseases as warmer air grows more charged with pollutants, mold spores, and pollens.

Global warming and climate change are not problems of any single region or area but these have global spread. As we know that the manner in which the concentration of greenhouse gases is increasing, in the atmosphere, is very harmful for sustenance of life on this earth. For protecting the human races and plant and animal species, rational measures are required to be framed and implemented. Sustainable utilization of the available natural resources, keeping the next generations in view, and minimizing the burning of fossil fuels, will help to reduce poisoning greenhouse gases. Use of non-conventional resources in lieu of conventional resources may sustain the life of non-renewable resources and can restore the environment. Equality in the process of development among the countries of the world, irrespective of their present status of development, will lead a way for maintaining sustainability and environmental restoration. The following measures can be taken for coping with the problems of global warming and climate change:

1. The total consumption of fossil fuels is increasing by about 1 per cent per annum. No steps currently being taken or under serious discussion are likely to prevent global warming in the near future. The challenge today is managing the probable effects while taking steps to prevent detrimental climate changes in the future.
2. Damage can be curbed locally in various ways. Coastlines can be armoured with dikes and barriers to block encroachments of the sea. Alternatively, governments can assist coastal populations in moving to higher ground. Some countries, such

as the United States, still have the chance to help plant and animal species survive by preserving habitat corridors, strips of relatively undeveloped land running north and south. Species can gradually shift their ranges along these corridors, moving towards cooler habitats.

3. There are two major approaches to slowing the build-up of greenhouse gases. The first is to keep carbon dioxide out of the atmosphere by storing it or its carbon component somewhere else – a strategy called carbon sequestration. The second major approach is to reduce the production of greenhouse gases.
4. The developed countries are all working to reduce greenhouse emissions. Several European countries impose heavy taxes on energy usage, designed partly to curb such emissions. Norway taxes industries according to the amount of carbon dioxide they emit. In the Netherlands, government and industry have negotiated agreements, aimed at increasing energy efficiency, promoting alternative energy sources, and cutting down greenhouse gas output.
5. International cooperation is required for the successful reduction of greenhouse gases. In 1992, at the Earth Summit in Rio de Janeiro, Brazil, 150 countries pledged to confront the problem of greenhouse gases and agreed to meet again to translate these good intentions into a binding treaty.
6. In 1997, in Japan, 160 nations drafted a much stronger agreement known as the Kyoto Protocol. This treaty, which has not yet been implemented, calls for the 38 industrialized countries that now release the most greenhouse gases to cut their emissions to levels 5 per cent below those of 1990. This reduction is to be achieved no later than 2012.
7. Large-scale afforestation should be taken in hand and at least 33 per cent of the landmasses should be covered by forests in Indian context.
8. Minimizing the use of non-renewable resources and development of renewable resources should be ensured.

CHAPTER 13

Environmental and Social Issues

Environmental and social issues are correlated and one of the prime concerns of the governments and societies worldwide. High level degradation of environment due to natural and man-induced activities impacted on societies and on the communities at a large scale. Tremendous exploitation of natural resources for the purpose of fulfilling human needs and also for development has resulted in vanishing of non-renewable resources. The growing need of natural resources and their high exploitation manifested a way for postulating the concept of 'sustainable development' which is the urgent need of the day. In all spheres of society, 'sustainable development' is becoming very important. Sustainable development is defined as the development that meets the needs of the present without compromising the ability of future generations to meet their own needs. It also looks at the equity between countries and continents, races and classes, gender and ages. It includes social development and economic opportunity on one hand and the requirements of environment, on the other. It is based on improving the quality of life for all, especially the poor and deprived, within the carrying capacity of the supporting ecosystems. It is a process which leads to a better quality of life while reducing the impact on the environment. Its strength is that it acknowledges the interdependence of human needs and environmental requirements. To ensure sustainable development, any activity that is expected to bring about economic growth must also consider its environmental impacts so that it is more consistent with long-term growth and development. According to The World Commission on Environment and Development (WCED), "sustainable development is a process of change in which the exploitation of resources, the direction of investments, the orientation of technological development, and institutional changes are made consistent with the future as well as present needs". The concept of sustainable development aims to bridge the gap between economic growth and environmental conservation. In developing countries, however, economic development and environmental conservation seem to remain as two diametrically opposed issues. The demands of the growing population for better standards of living drive nations in the developing world to exploit resources, endanger the environment and push it to the brink of collapse. Sustainable development is the key concept that provides an answer to how we can co-exist with nature harmoniously, maintaining a delicate balance between the human needs to improve lifestyles and attain a feeling of well-being on one hand, and the need to preserve the natural resources and ecosystems on which we and future generations depend, on the other. Focussing on the four key points of the WCED's definition of sustainable development, we in the developing world should work for a stable relationship between human activities and the natural world, which does not diminish the prospects for future generations to enjoy a quality of life as good as, if not better, than our own today. The other environmental and social issue is environmental pollution. Air, water, soil and noise are the major forms of pollution. These pollutions together impacted on the health quality and the environment becoming polluted and degraded largely.

Environmental Issues

As cited above, uneven development and pollution are the major environmental and social issues. Starting with pollution – air, water, noise and soil – it is a major social and environmental problem. Polluting industries, including thermal power plants, have been a significant source of air and water pollutions. Monitoring of designated cities and towns for air pollution revealed that while the

Box 13.1 Initiatives to Control Environmental Pollution

- Existing policy initiatives to improve environment like the National Conservation Strategy and Policy Statement for Environment and Development, 1992, Policy Statement for abatement for Pollution, 1992 and National Policy, 1988.
- Stipulation of ambient and industry specific emission and effluent standards.
- Setting up of clean technology mechanisms in polluting industries.
- Setting up of Common Effluent Treatment Plants (CETPs) in industrial estates.
- Establishing waste minimization circles (WMC) in clusters of small scale industries.
- Implementing recommendations of Charter of Corporate Responsibility for Environmental Protection (CREP) in 17 categories of highly polluting industries.
- Prior environmental clearance of development projects based on impact assessment.
- Implementation of an eco-mark scheme to encourage production/consumption of environment – friendly products.
- For controlling vehicular pollution, progressive emission norms at the manufacturing stage have been notified, cleaner fuels like unleaded petrol, low sulphur diesel and compressed natural gas (CNG) introduced.
- Promotion of economic instruments to internalize the costs of pollution and fiscal incentives for pollution control equipment.
- Monitoring of ambient air and water quality, bio-monitoring of rivers/lakes and identification of hazardous wastes streams in various sectors/processes.

levels of sulphur dioxide were within the standard limits, those of oxides of nitrogen exceeded the limits in many cities, with high levels of suspended particle matter (SPM) being the more prevalent form of air pollution in almost all the metros. Similarly, a recent assessment of water quality of aquatic resources reveals that while organic pollution continued to be the predominant pollutant, there is a gradual improvement in this indicator of water quality over time. Flyash, phospho-gypsum and iron and steel slags comprise the main forms of solid wastes generated in India.

The challenge of sustainable development requires integration of the quest for economic development with environmental concerns. Traditional energy and resource intensive and waste-generating models of industrialization not only impose a heavy cost on society, in terms of pollution and deteriorating quality of life, but also are unsustainable in the medium to long run, particularly with exhaustible natural resources. Government has put in place necessary legislative and regulatory measures, both preventive and promotive, for protection, conservation and development of the environment.

Box 13.2 New Policy Initiatives

- Formulation of a National Environment Policy.
- Setting up of National Clean Development Mechanism (CDM) Authority as per Kyoto Protocol.
- Re-engineering of environmental clearance process with a view to bringing greater transparency and efficiency in the clearance process.
- Revisiting the Coastal Regulation Zone (CRZ) Notification to enable environmentally sustainable use of coastal resources.
- Developing a National Chemical Management Profile for the country.

Issues in Environment Policy

Consumption and Conservation of Natural Resources

Sustainable development goes beyond a static framework of correcting environmental externalities, deficient markets and inadequate property rights. It is very much concerned with the availability of environmental and natural resources in the future. The problem may be particularly acute in developing countries that have high discount rates due to the scarcity of capital and prevalence of poverty. Therefore, a rational discount rate is central to maintaining an optimal balance between consumption and conservation of natural resources.

Valuation of Environmental Damages

Attempts to establish optimal pollution abatement levels or to set more general environmental quality objectives confront several problems in determining the monetary value of environmental damage. One difficulty in setting optimal abatement levels is that an economic activity must be functionally related to measurable environmental changes and the scientific basis for establishing the linkages may not exist. Secondly, only a use value of a good is normally accepted as contributing to welfare. Environmental goods have non-use value and option value, also. Thirdly, many

Table 13.1 Estimated Cost of Environment Degradation in Selected Asian Countries

Country	*Year or Period*	*Environmental Damage*	*Annual Cost (US$ billion)*	*Cost as % of GDP (%)*
China	1990	Productivity losses caused by soil erosion, deforestation and land degradation, water shortage and destruction of wetlands. Health and productivity losses caused by environmental pollution in cities.	13.9–26.6	3.8–7.31.7
Indonesia	1989	Health effects of particulate and lead levels above WHO standards in Jakarta.	6.3–9.3 2.2	2.5 2.0
Pakistan	Early 1990s	Health impacts of air and water pollution and productivity losses from deforestation and soil erosion.	1.7	3.3
Philippines	Early 1990s	Health and productivity losses from air and water pollution in the vicinity of Manila.	0.3–0.4 1.6	0.8–1.0 2.0
Thailand	1989	Health effects of particulate and lead levels above WHO standards.		
India	1992	Costs of urban air pollution, water pollution, soil erosion, land degradation and deforestation.	10–13.8	4.5–6.0

Source: Human Development Report, 1998, pp. 67, 79.

environmental services do not pass through markets and, therefore, no market prices exist to indicate value. Determining the value also depends on the type of environmental damage under consideration. Among the economic valuation techniques adopted for such valuation of environmental damage are included the contingent valuation method, travel cost method, hedonic pricing, surrogate markets method, etc. The importance of valuation of environmental damage lies in demonstrating the existence of substantial economic benefits from increased resource protection. An appropriate emphasis, therefore, needs to be placed on valuation of such environmental damages and its suitable incorporation in the development projects.

Natural Resource Accounting

Accounting schemes that more accurately measure the environmental costs of economic activity at all levels could help increase the efficiency of natural resource use and reduce related environmental impacts. The need for Natural Resource Accounting (NRA) and their integration with the system of national accounts has been emphasised in various policy documents of the Ministry of Environment and Forests (MoEF). The Ministry in 1993 had developed a framework for preparing such integrated accounts for India. It has further taken up programme for development of prototype accounts for air, water, biodiversity and common property resources for selected hotspots in the country with a view to firm up the methodologies and techniques in their preparation.

Use of Economic Instruments

While regulatory measures remain essential, new approaches for considering market choices in the protection of environment are being increasingly adopted by both developed and the developing countries. The aim is to give industries and consumers clear signals about the cost of using environmental and natural resources. If the environmental costs of economic activity are more adequately reflected in the prices paid for goods and services, then companies and ultimately the consumers would be guided to adjust their market behaviour so as to reduce pollution and waste. The expectation is that price signals will influence behaviour to avoid excessive use of natural resources. Hence, a greater use of such market-based approaches as a complement to the present regulatory system is desirable. Use of such incentives would facilitate operationalizing the standard approach of 'Polluter Pays Principle'.

Removing Subsidies that Encourage Unsustainable Use

Production of natural resource-related commodities is often subsidised. For example, power tariff charged is only a fraction of costs of generating power. Such subsidies encourage wasteful use of resources, often creating or exacerbating environmental problems. Salination from over-irrigation is a major cause of land degradation in the country. Imbalance in the use of fertilizer nutrients, especially owing to underpricing such as urea, affects soil productivity adversely. Low price of kerosene has encouraged its adulteration especially of diesel and motor spirit contributing to air pollution. Similarly, the water and sanitation problem may be largely due to low user charges, starving the implementing agencies of funds to maintain and operate these facilities. Eliminating or reducing such subsidies would confer both economic and environmental advantages. However, in certain cases, elimination of such subsidies may require weeding out of inefficient and marginal enterprises, thus raising unemployment. A careful targeting of subsidies where these are most needed is, therefore, called for.

Extension of Property Rights

In some cases, the problem of environmental externality derives not from lack of appropriate user charges but from the lack of or ill-defined property rights over resources. The extensive degradation of village commons or overexploitation of lakes and oceans for fisheries are the examples of such externalities. In such cases it may be more efficient for the government to assign or clarify property rights and allow private agents to handle problems of environmental quality through negotiations among the affected parties.

Trade and Environment

The concern with environmental implications of trade involves both the domestic implications of policy reforms as well as the global environmental dimension of international trade agreements.

Domestic implications of trade policy would suggest growth of less polluting industries and shift to higher quality and environment-friendly products/processes for traditional polluting industries (like leather and textiles) through adoption of cleaner technologies, so that greater efficiency and productivity associated with liberalized trade policy may also reduce pollution intensity. As regards the global environmental dimension of international trade, the debate has revolved around the issue of whether free trade is beneficial to global and national environmental conditions and whether it should be used to influence national and international environmental standards and agreements. A number of countries have adopted trade-related environmental measures with implications for product/process standards, packaging and labelling of export merchandise. These measures have the impact of non-tariff barriers and need to be tackled through improved product/process quality. Removing such trade barriers would increase employment, promote economic diversification in developing countries, and facilitate a greater effort towards improvement of the environment.

Box 13.3 Green Accounting

- National accounts have been providing the most widely used indicators for the assessment of economic performance, trends in economic growth and the economic counterpart of social welfare. However, the new emphasis on sustainable development draws attention to the need for a broader assessment of growth and welfare by modified national accounts. In assessing cost and capital, national accounts do not consider scarcities of natural resources which threaten the sustained productivity of the economy and the degradation of environmental quality and consequential effects on human health and welfare. In addition, some expenditure for maintaining environmental quality is accounted as increases in national income and product. This is despite the fact that such outlays could be considered a maintenance cost of the society, rather than social progress. Thus, the conventional accounts are likely to send wrong signals and may result in

Cont'd...

...Cont'd

policy decisions which are non-sustainable for the country. Green accounting, on the other hand, is focused on addressing such deficiencies in conventional accounts with respect to the environment.

- Integrated environment and economic (green) accounting, therefore, attempts at accounting for both socio-economic performance and its environmental effects and integrating environmental concerns into mainstream economic planning and policies. Such accounting implies allocating environmental costs (and benefits) to those activities and sectors that have caused them, in other words accounting for accountability, is a prerequisite for national management of both the environment and the country. Given the experimental nature of some of the proposed methodologies, particularly those on monetary valuation of non-marketed asset and externalities, such environmental accounting would require numerous and controversial estimates and valuation. Therefore, rather than modifying the 'core' systems of SNA, their incorporation through a system of 'satellite accounts' have been suggested.
- Such integrated accounts can be useful in assessing the sustainability of economic growth and also the structural distortion of the economy by environmentally unsound production and consumption patterns. However, lack of international consensus on how to incorporate environmental assets and costs (and benefits) of their use in national accounts and existence of low statistical capacities for measuring natural resources depletion and environmental quality changes have resulted in a slow progress in development of green accounts.
- Nevertheless, the idea of placing statistical coverage of environmental concerns in a national accounts framework commands widespread support. Already, several attempts have been made at experimenting with satellite accounts – notably in Costa Rica, Mexico, the Netherlands, Norway and Papua New Guinea, among others. Indicative estimates suggest that conventionally measured GDP may exceed GDP adjusted for natural resources depletion and environmental degradation by between 1.5 per cent and 10 per cent.

Development that can Reduce Poverty

Efforts should be made to encourage economic growth and provide employment to the poor. The growth strategy should also accord weight to the direct intervention programmes for poverty alleviation, stabilization of population growth rate and greater employment opportunities for both rural and urban poor. Further, it should target human capital investment in health care and education, provisions of basic civic amenities like drinking water and sanitation, legal reforms to extend land tenure and other rights to the rural poor and enhancing the status of women while increasing their participation in development.

Box 13.4 Environmental Taxes

- As against the command and control approach to management of the environment, the Economic or Market Based Instruments (MBIs) approach sends economic signals to the polluters to modify their behaviour. The approach normally involves financial transfers between polluters and the community and affects relative prices. The approach based on standards and regulations. The MBIs, therefore, have the benefit of being flexible and cost effective providing incentives for dynamic efficiency and resource transfer.
- Economic instruments used for environmental tax include pollution charges (emission/affluent tax/pollution taxes), marketable permits, deposit refund system, input taxes/product charges, differential tax rates and user administrative charges and subsidies for pollution abatement. These can be both price-based and quality-based instruments.
- MBIs have been applied in both developed as well as developing countries. In general, price-based MBIs have been more widely used than those which are quantity-based. Within price-based MBIs, indirect instruments such as input-output taxes, differential tax rates and user fees have found extensive application in developed countries. By contrast, developing countries have made greater use of subsidies including those for end-of-pipe treatment equipment.

Cont'd...

...Cont'd

- The main MBIs used in India are subsidies for pollution abatement equipment for air and water resources. This provides rebates on duties for various pollution control equipment, monitoring instruments and abatement machinery for air/water pollution and promotion of unleaded fuel/fuel efficient automobile subsidy on automobile pollution kits/converters, etc. Accelerated depreciation for pollution control machinery is also provided. Among user charges/administration charges, consent fee is charged from industries under the Water Act and the Air Act. A water cess based on the consumption of water and type of industry (polluting) is also levied on selected industries and urban municipalities to conserve consumption and control pollution of water.

People's Participation

People's participation is an essential ingredient of environment management along with other components like regulation and use of economic instruments. People's participation is facilitated by various environmental education and awareness programmes. The MoEF implements formal as well as informal education programmes and also conducts various environmental awareness programmes. The usefulness of public participation is highlighted by the success of programmes like the Joint Forestry Management (JFM). A common approach followed by many countries, including India, towards greater public participation is the scheme of eco-labelling, which helps consumers to identify products that are environment-friendly. Public hearing of major projects coming for environmental clearance to the MoEF has also been made mandatory. Other approaches in practice include the "consumer cooperatives" in Japan which popularize green products which are recyclable, biodegradable, rechargeable, ozone-friendly and unleaded. The Indonesian experiment with rating and public disclosure of the environmental performance of industries/factories has also vindicated the effectiveness of public participation/opinion in management of the environment.

Participation in Global Dimensions of Environment

India is a signatory to various international conventions having a direct bearing on environmental protection and conservation. Such conventions play an important role in harmonizing approaches and ensuring joint action on various global environmental issues. They also facilitate flow of larger funds and greater access to new technologies, contribute to protection of natural resource commodities and promote more sustainable production methods. Environmental protection and conservation of natural resources emerged as the key priority in the wake of the Stockholm Conference on Human Environment in 1972. India is also a party to the Rio Declaration on Environment and Development and the Agenda 21 – the operational programme for sustainable development.

The United Nations Framework Convention on Climate Change (UNFCCC) aims at stabilization of greenhouse gas concentrations in the atmosphere at levels that would prevent dangerous anthropogenic interference with the climate system. It enjoins upon the parties to implement commitments contained in the various provisions of the Convention. As per the existing commitments, India is not required to adopt any reductions and limitation of greenhouse gas emissions. However, the third Conference of Parties to the Convention, held in Kyoto, Japan in December 1997, considered an alternate protocol for strengthening of commitments of the parties to achieve the objectives of the Convention. The Kyoto Conference witnessed deliberate efforts of the developed countries to introduce new commitments for developing countries on the one hand, and to secure flexibility to implement their commitment through new mechanisms (such as emission trading, bank borrowing and joint implementation), on the other. The final agreement calls upon for an average cut in emissions of greenhouse gases to 5.2 per cent below 1990 levels to be achieved between 2008 and 2012 (European Union 8 per cent reduction, Japan 6 per cent reduction and USA 7 per cent reduction). The provision of joint implementation and emission trading, however, would be confined to developed countries only. A clean development mechanism has also been defined for use by both developing and developed countries for sustainable development and implementation of commitments respectively.

The Convention on Biological Diversity attempted at conservation of biodiversity, sustainable use of its components and fair and equitable sharing of the benefits arising out of utilization of genetic resources. Several steps, like preparation of a National Action Plan, framing legislation on biodiversity, intellectual property rights on patenting of micro-organisms, etc., have been initiated to meet our commitments under the Convention, as also to restructure administrative and policy regime in tune with the Articles of the Convention. Similarly, by acceding to the Convention to Combat Desertification, India has also reiterated its commitment to continue to accord priority to all actions to prevent land degradation and to improve the productivity of land on which the poor depend for their subsistence in such areas. With a view to strengthen the global efforts to protect the ozone layer, India acceded to the Montreal Protocol in 1992. Draft rules on Ozone Depleting Substances (ODS) phase out under the Environment (Protection) Act, 1986 have been prepared after extensive consultations with the industry, NGOs and concerned government departments. The policy to issue licences for import of ODS has since been formulated. Duty exemptions are granted for new investment with non-ODS technologies. Instructions have been issued to all commercial banks prohibiting finance/refinance of new investments with ODS technologies.

Financing Sustainable Development

The share of environment and forests sector in the Eighth Five-Year Plan outlay amounted to only 1.1 per cent, as compared to other social sector shares of 1.2 per cent for urban development, 1.3 per cent for housing, 1.4 per cent for family welfare, 1.6 per cent for medical and public health and 3.9 per cent for education. About 75 per cent of this approved outlay on environment and forests was in the state sector with bulk of the investment being in the forestry sector. Keeping in view the resource constraints, the MoEF has identified in its Environment Action Programme priority areas covering conservation of and sustainable utilisation of biodiversity in selected ecosystems; afforestation, wasteland development, conservation of soil and moistures including measures to lessen/eliminate water pollution; controlling industrial and related pollution with accent on reducing and/or management of wastes particularly those of hazardous nature; improving access to cleaner technologies; tackling of urban environmental issues; and strengthening the scientific understanding of environmental issues as well as structures at different levels and orientation and creation of environmental awareness and an alternative energy plan with emphasis on cleaner sources of fuel.

Table 13.2 Approved Outlays: Environment and Forests Sector in India

	VII Plan	*VIII Plan (Rs crore)*
Ecology and Environment	428	828
Forests and Wildlife	1859	4082
Total, Environment and Forests	2287	4910
(a) Central sector	797	1200
(b) State sector	1490	3710

Source: MoEF.

Towards Sustainable Development

Until two decades ago the world looked at economic status alone as a measure of human development. Thus, countries that were economically well developed and where people were relatively richer were called advanced or developed nations while the rest where poverty was widespread and were economically backward were called developing countries. Most countries of North America and Europe, which had become industrialized at an earlier stage, have become economically more advanced. They not only exploited their own natural resources rapidly but also of developing countries to grow even larger economies. Thus, the way development progressed, the rich countries got richer while the poor nations got poorer. However, even the developed world has begun to realize that their lives were being seriously affected by the environmental consequences of development based on economic growth alone. This form of development did not add to the quality of life as the environmental conditions had begun to deteriorate. By the 1970s most development specialists started to appreciate the fact that economic growth alone could not bring about a better way of life for people unless environmental conditions were improved.

Development strategies, in which only economic considerations were used, had begun to suffer from serious environmental problems caused due to air and water pollution, waste management, deforestation and a variety of other ill-effects that seriously affected people's well-being and health. There were also serious equity issues between the haves and the have-nots in society, both at the global and national levels. The disparity between the lifestyles of the rich and the poor was made worse by these unsustainable development strategies. Many decades ago, Mahatma Gandhi envisioned a reformed village community based on sound environmental management. He stressed the need for sanitation based on recycling human and animal manure and well-ventilated cottages built of recyclable material. He envisaged roads as being clean and free of dust. His main objective was to use village-made goods instead of industrial products. All these principles are now considered part of sound long-term development. Gandhiji had designed a sustainable lifestyle for himself when these concepts were not a part of general thinking. A growing realization of the development strategy that he had suggested many decades earlier is now accepted by experts on development across the world. This is based on his concept that the world could support people's needs but not their greed. It has become obvious that the quality of human life has worsened as economies grew. The world now appears to be at a crossroads. It has taken the path of short-term economic growth and now suffers the consequences of environmental degradation at the cost of loss of 'quality of human life'. The earth cannot supply the amount of resources used and wasted by the economically well-off sectors of society as well as that required for day-to-day sustenance of the ever-growing population in less developed countries. Society must thus change its unsustainable development strategy to a new form where development will not destroy the environment. This form of sustainable development can only be brought about if each individual practices a sustainable lifestyle based on caring for the earth. It was also realized that these were not simple issues. At Stockholm Conference in 1972 Indira Gandhi said that poverty was the greatest polluter. This meant that while the super rich nations had serious environmental problems, the underdeveloped in Asia, Africa and South America had a different set of environmental problems linked to poverty. Developing countries were suffering the consequences of a rapidly expanding human population with all its effects on the overutilization of natural resources. Thus, increasingly, the world began to see the need for a more equitable use of earth resources. The control over natural resources and the wealth that it produces also begins to create tensions between people that can eventually lead to both strife within a country and wars between nations. This is also a major cause for the loss of quality of life. How then could a new form of development be brought about that could solve the growing discontent in the world?

It has become obvious that development must begin to change from aiming at short-term economic gains to a long-term sustainable growth that would not only support the well-being and quality of life of all people living in the world today but that of future generations as well. The current strategies of economic development are using up resources of the world so rapidly that our future generations, the young people of the world, would have serious environmental problems, much worse than those that we are facing at present. Current development strategies have now come to be considered unsustainable for the world's long-term development. The newer concept of development has come to be known as 'sustainable development'. The nations of the world came to clearly understand these issues at the Rio Conference in 1992. Several documents were created for the United Nations Conference on Environment and Development (UNCED), which brought out the fact that environment and development were closely connected and that there was a need to 'care for the earth'.

Many development projects – dams, mines, roads, industries and tourism development – have severe environmental consequences that must be studied before they are even begun. Thus, for every project, there must be a scientifically and honestly done Environmental Impact Assessment (EIA), without which the project must not be cleared. Large dams, major highways, mining, industry, etc., can seriously damage ecosystems that support the ecological health of a region. Forests are essential for maintaining renewable resources, reducing carbon dioxide levels and maintaining oxygen levels in the earth's atmosphere. Their loss impairs future human development. Loss of forests depletes biodiversity which has to be preserved to maintain life on earth. Major heavy industries, if not planned carefully, lead to environmental degradation due to air and water pollution and generate enormous quantities of waste that lead to long-term environmental hazards. Toxic and nuclear wastes can become serious economic problems as getting rid of them is extremely costly. Thus, the economic benefits of a project must be weighed against the possible environmental costs before a project is permitted. We, as citizens of our nation, and increasingly as citizens of one common future at the global level, must constantly monitor the pattern of development. If we see that a development project or an industry is leading to serious environmental problems, it is our duty to bring this to the attention of authorities such as the local administration, the Forest Department or the Pollution Control Board, to look into the issue. Further, if new development projects are being planned in and around the place where we live it is our duty to see that this is brought about in accordance with environmental safeguards. While we all need to think globally, we need to act locally. We have to see that we change development from its present mandate of rapid economic growth without a thought for future ecological integrity, to a more sustainable ecologically appropriate strategy. If new projects of a large size are to be passed the government has made it compulsory to publish the summary report of the EIA and conduct a 'public hearing'. It is essential that all of us as responsible citizens read, evaluate and respond to such public hearings held in our area and make comments on the possible impacts of the project. In many situations there are proponents of the project who only look at their own rapid economic gains. It is for citizens as concerned individuals and groups to counter these vested interests so that our environment is not degraded further. Life has to be made more livable for all. We cannot support the economic growth of one sector of society while we permit environmental degradation to destroy the lives of the less fortunate.

Major Energy Problems

Urban centres use enormous quantities of energy. In the past, urban housing required relatively smaller amounts of energy than we use at present. Traditional housing in India required very little temperature adjustments as the materials used, such as wood and bricks, handled temperature changes better than the current concrete, glass and steel of ultra-modern buildings. Until the 1950s many urban kitchens were based on fuelwood or charcoal. This was possible and practical when homes had chimneys and kitchens were isolated from the rest of the house. Smoke became a problem once this changed to apartment blocks. Kerosene thus became a popular urban fuel. This changed to electrical energy and increasingly to natural gas by the 1970s in most parts of urban India. Urban centres in hot climates need energy for cooling. The early systems of fans changed into air-conditioning, which consumes enormous quantities of energy. New buildings in our country have been using large areas covered by glass. While in cold climates this uses the greenhouse effect to trap the warmth of the sun inside, in the hot climate this adds several degrees to the temperature inside. Thus, it requires even more energy to run large central air-conditioning units. High-rise buildings in urban centres also depend on energy to operate lifts and an enormous number of lights. Urban transport depends on energy mainly from fossil fuels. Most urban people use their own individual transport rather than public transport systems for a variety of reasons. Urban transport in different cities and even different parts of a city is either inefficient or overcrowded. Thus, even middle income groups tend to use their own private vehicles. This means more and more vehicles on the road which leads to traffic congestion, waste of time for all the commuters, and a great load of particulate matter and carbon monoxide from the exhaust of vehicles. This causes a rise in the number of people having serious respiratory diseases. Thus, there is a need to develop a more efficient public transport system and discourage the use of individual vehicles. Each of us as an environmentally conscious individual must reduce the use of energy. An unnecessary light left on carelessly adds to energy use. Imagine the amount of energy wasted by thousands of careless people. If we learned to save electricity, we would begin to have a more sustainable lifestyle.

Resettlement and Rehabilitation of Project-Affected People

Major projects such as dams, mines, expressways, or the notification of a national park disrupts the lives of the people who live there and may also require moving them to an alternative site. None of us would like to give up the home we grew up in. Uprooting people is a serious issue. It reduces their ability to subsist on their traditional natural resource base and also creates great psychological pressures. Especially tribal people, whose lives are woven closely around their own natural resources, cannot adapt to a new way of life in a new place. Thus, no major project that is likely to displace people can be carried out without the consent of the local people. In India, lakhs of people have been unfairly displaced by thousands of dams built since independence to drive the green revolution. The dams have been built virtually at the cost of these poor local people who have been powerless to resist the government's will. The government is expected to find 'good' arable land to resettle displaced persons and provide them with an adequate rehabilitation package to recover from the disruption. This has rarely occurred to the satisfaction of the project-affected individuals. In many cases across the country, this has not been implemented satisfactorily for decades.

Resettlement requires alternate land. However, in our overpopulated country, there is no arable high quality land available. Thus, most project-affected persons are given unusable wasteland. Rehabilitation involves more than just giving land. In most cases this is also not adequately done. The greatest battle to save their own precious land has been carried out by the tribal people of the Narmada river area. They have fought to save their lands for decades. The Narmada Bachao Andolan has shown how bitter people can get over this issue. Resettlement not only puts pressure on the project-affected people but also on the people who have been living in the area that has been selected for resettlement. Thus, both the communities suffer and conflict over resources is a distinct possibility in future. There are however situations where communities request for shifting to a new site. This is often observed where people live inside or on the periphery of a national park or wildlife sanctuary. In these situations, such as the Gir in Gujarat, the local people have asked to be given alternate land where they could live peacefully away from lions that kill their cattle, but the government has been unable to find suitable areas where they can be shifted for decades.

Environmental Ethics: Issues and Possible Solutions

Environmental ethics deal with issues related to the rights of individuals that are fundamental to life and well-being. These concern not only the present generation, but also future generations. These also deal with the rights of other living creatures that inhabit our earth.

Resource Consumption and Optimum Utilization

Environmental ethics deal with issues that are related to how we utilize and distribute our resources. Can individuals justifiably use resources so differently that one individual uses resources many times more lavishly than other individuals who have barely enough to survive? In a just world, there has to be a more equitable sharing of resources than we encounter at present. The just distribution of resources has global, national and local concerns that we need to address. There are rich and poor nations. There are rich and poor communities in every country. And, there are rich and poor families. In this era of modern economic development, the disparity between the haves and have-nots is widening. Our human environments in the urban, rural and wilderness sectors use natural resources that shift from the wilderness (forests, grasslands and wetlands) to the rural sector and from there to the urban sector. Wealth also shifts in the same direction. This unequal distribution of wealth and access to land and its resources is a serious environmental concern. An equitable sharing of resources forms the basis of sustainable development for urban, rural and wilderness dwelling communities. As the political power base is in the urban centres, this itself leads to inequalities and a subsequent loss of sustainability in resource management in the rural and even more so for forest dwelling people.

In 1985, Anil Agarwal published the first report on the *Status of India's Environment*. It emphasized that India's environmental problems were caused by the excessive consumption patterns of the rich that left the poor poorer. It was appreciated for the first time that tribal, especially women and other marginalized sections of our society, were being left out of economic development. There are multiple stakeholders in Indian society who are dependent on different natural resources which cater directly or indirectly to their survival needs. Agarwal brought forth a set of eight propositions which are of great relevance to the ethical issues that are related to environmental concerns. These include:

- Environmental destruction is largely caused by the consumption of the rich.
- The worst sufferers of environmental destruction are the poor.
- Even where nature is being 'recreated', as in afforestation, it is being transformed away from the needs of the poor and towards those of the rich.
- Even among the poor, the worst sufferers are the marginalized sections, cultures and occupations, and most of all, women.
- There cannot be proper economic and social development without a holistic understanding of society and nature.
- If we care for the poor, we cannot allow the Gross Nature Product to be destroyed any further. Conserving and recreating nature has become our highest priority.
- Gross Nature Product will be enhanced only if we can arrest and reverse the growing alienation between the people and the common property resources. In this we will have to learn a lot from our traditional cultures.
- It is totally inadequate to talk only of sustainable rural development, as the World Conservation Strategy does. We cannot save the rural environment or rural people dependent on it, unless we can bring about sustainable urban development.

  Who pays for the cost of environmental degradation? Most sections of society do not feel the direct effects of degradation of the environment till it is too late. Those who suffer most are the poor, especially rural women, and tribal people who are dependent on forests. Traditional fishermen who are dependent on streams and rivers, and coastal people who fish and catch crustacea, are seriously affected by the degradation of aquatic ecosystems. Fuelwood gatherers from different types of forests, and pastoralists who are dependent on common grazing lands suffer when their resources are depleted.

Several marginalized sections of society are most affected by deforestation, or the loss of grassland tracts, or the deterioration of perennial water sources. All these effects can be linked to unsustainable increasing pressures on land and natural resources. The well-to-do educated urban dweller consumes much larger quantities of resources and energy than the traditional rural individual. Urban dwellers who are far removed from the source of natural resources that sustain their lives thus require exposure to a well-designed environment education programme to appreciate these issues. While the rural people have a deep insight into the need for sustainable use of natural resources and know about methods of conservation, there are however several newer environmental concerns that are frequently outside their sphere of life experiences. Their traditional knowledge of environmental concerns cannot be expected to bring about an understanding of issues such as global warming, or problems created by pollution, pesticides, etc. These people thus require a different pattern of environmental education that is related to their gaps in information. With the rapidly changing rural scenario the development that is thrust on unsuspecting rural communities needs to be addressed through locale-specific environment awareness programmes designed specifically for rural school children and adults. This must also use their local traditional knowledge systems as a base on which modern concepts can be built, rather than by fostering concepts that are completely alien to their own knowledge systems.

Common property resources (CPRs) in India once included vast stretches of forests, grazing lands and aquatic ecosystems. When the British found that they were unable to get enough wood for ship building and other uses, they converted forest areas into government 'reserved forests' for their own use to grow timber trees. This alienated local people from having a stake in preserving these resources. This in turn led to large-scale losses in forest cover and the creation of wasteland. In the past, in traditional villages that were managed by local panchayats, there were well-defined rules about managing grazing lands, collecting forest resources, protecting sacred groves, etc., that supported conservation. There was a more or less equitable distribution that was controlled by traditional mechanisms to prevent misuse of CPRs. Any infringement was quickly dealt with by the panchayat and the offender was punished. CPRs were thus locally protected by communities. As land use patterns changed, these mechanisms were lost and unsustainable practices evolved, frequently as a result of an inadequately planned development strategy.

Wasteland Reclamation and Development

Loss of vegetation cover leads to loss of soil through erosion, which ultimately creates wastelands. This is one of the pressing problems of the country. Loss of soil has already ruined a large amount of cultivable land in our country. If it remains unchecked, it will affect the remaining land. Unless we adequately safeguard our 'good' lands, we may eventually face a serious shortage of food grains, vegetables, fruits, fodder and fuelwood. Hence, conservation of soil, protecting the existing cultivable land and reclaiming the already depleted wastelands figure prominently among the priority tasks of planning for the future. Some of the wasteland reclamation programmes have been unsuccessful because after sometime the land reverts to its original poor condition due to mismanagement and unscientific ways in which the reclamation has been carried out. In choosing wasteland reclamation methods attention must be paid to the cost factor. This has to be taken into account before deciding on a particular method for reclamation of wastelands. Proper studies of environmental aspects and human impacts, which are responsible for the development of wastelands, have to be looked into. Wasteland can be classified into three forms: (1) easily reclaimable, (2) reclaimable with some difficulty and (3) reclaimable with extreme difficulty. Easily reclaimable wastelands can be used for agricultural purposes. Those which can be reclaimed with some difficulty can be utilized for agro-forestry. Wastelands that are reclaimed with extreme difficulty can be used for forestry or to recreate natural ecosystems.

Agriculture: Wasteland can be reclaimed for agriculture by reducing the salt content which can be done by leaching and flushing. Gypsum, urea, potash and compost are added before planting crops in such areas.

Agro-forestry: This involves putting land to multiple uses. Its main purpose is to have trees and crops inter- and/or underplanted to form an integrated system of biological production within a certain area. Thus, agro-forestry implies integration of trees with agricultural crops or livestock management simultaneously.

Forestry: Attempts to grow trees in highly non-alkaline saline soils have been largely unsuccessful. Field experiments have shown that species like Eucalyptus, Prosopis and Acacia Nilotica could not be grown in highly alkaline soil. Studies have shown that if tree seedlings are planted with a mixture of original soil, gypsum and manure, better growth can be achieved. It is however important to use indigenous species of trees so that the programme recreates the local ecosystem with its entire species.

Wasteland development provides a source of income for the rural poor. It ensures a constant supply of fuel, fodder and timber for local use. It makes the soil fertile by preventing soil erosion and conserving moisture. The programme helps maintain an ecological balance in the area. The increasing forest cover helps in maintaining local climatic conditions. Regenerated vegetation cover helps in attracting birds which feed on pests in the surrounding fields and function as natural pest controllers. The trees help in holding back moisture and reduce surface run off rates thus helping in the control of soil erosion.

The first major task is the identification of the problem at the micro level. For this it is necessary to have district, village and plot level surveys of the wasteland. A profile of the maps indicating the detailed distribution and information on the wasteland is essential. With the help of local government institutions such as the village panchayats, along with Block Development Officers, and Revenue Department functionaries, a plan based on the community needs must be produced. This must be done through a participatory exercise that involves all the different stakeholders in the community. A think-tank of administrators, ecologists and local NGOs must also be involved in the process. The next step is to identify the factors that are responsible for the formation of wasteland. Based on these factors the wasteland is classified into marginally, partially or severely deteriorated lands. Locale specific strategies for reclaiming the wasteland must be worked out. Government officials along with the local NGOs must assist the farmers by demonstrating improved methods of cultivation, arranging for loans for the small, marginal and landless farmers and the people from the weaker sections of the society. Involving local women has proved to be of great value. Another essential component of the programme is to organize publicity campaigns, integrated with training farmers and frontline government and Forest Department staff on the various aspects of wasteland utilization. Environmental scientists can help by suggesting the necessary changes in cropping patterns, particularly for drought-prone areas. Other tasks that should be addressed include the selection of appropriate crops for fodder and trees that provide local people with non-wooded forest products according to the nature of the wasteland. Testing soil in laboratories provides guidance to the farmers on the proper land management to be used. Irrigation and other expertise needed for improving productivity without creating unsustainable patterns of development provide the local people with newer technological advances. Guidelines regarding control of waterlogging must be provided. Appropriate technologies must be made available to people belonging to the weaker sections and landless farmers. Collective efforts have to be made to check soil losses through water and wind erosion to prevent the collapse of the irrigation system through siltation. Plans concerning wasteland reclamation and utilization prepared at various stages must be properly integrated for a successful long-term outcome. The demands of our increasing human population for environmental goods and services have imposed severe pressures on the available land resources, especially on the forests and green cover. This is closely linked to the well-being of the rural population which constitutes a large proportion of the population depending on local natural resources for their survival. The increasing demand for food, fodder, fuelwood, etc., has led to activities that are responsible for increasing environmental degradation which, in turn is responsible for the extension of wastelands. Thus, the development of agro-forestry based agriculture and forestry has become the prime prerequisite for an overall development of the economy in the country. The pressure on land is already very high and the only hope of increasing productivity lies in bringing appropriate improvement in the various categories of wasteland spread over the country.

Refuse, Reduce, Reuse, Recycle and Recover

Refuse, Reduce, Reuse, Recycle and Recover, or the 5Rs principle, is the new concept in waste management. But what does it actually mean? Although some waste is inevitable in any society, we must minimize the generation of waste at the source by using minimal resources. Do not use what you do not need. The goal of every society should be to reach a low-waste or no-waste society. Use your own reusable cloth/jute bags instead of plastic bags. The residual waste can be converted into a useable resource. In developed countries waste is used to produce energy. Several technological breakthroughs have recently been made to recover material from industrial waste like heavy metals and chemicals such as mercury and nitric acid. Thus, the waste does not remain a waste product anymore, but becomes a useful resource; e.g., using kitchen wet waste to make compost that can be used as an organic fertilizer, using sewage in a biogas plant to make fuel, one industry's waste could be a valuable resource for another industry, cloth rags from the textile industry are bought and used by paper and other industries. The 3Rs principle of Reduce, Reuse and Recycle should be followed in that order.

- *Reduction* is the best option. If we reduce at source, there is a smaller chance of waste generation and the pressure on our already stretched natural resources is reduced.
- *Reuse* is the next best option, as the product is reused in its current form without any energy expended to convert it into a new item.
- *Recycling* is the last option, as although it converts a waste into a resource, it uses energy to transform that resource into a new useable product.

Thus, by following the 'Reduce, Reuse and Recycle' principle, i.e., by reducing use at source, by reusing and recycling whatever possible and finally by proper disposal of residual waste, we can cut down or the waste generated and ensure that the minimal residual waste does not harm our environment. This principle can be followed by everyone – from an individual or an industry to a whole country. What can I do? You can follow the 3Rs principle in the following ways:

- Use only as much as you need, be it any resource – water, food, paper, etc.
- Next time you throw away something, think about whether it is really a waste. If it is of no use to you, could someone else use it? Reuse rinses water to water your garden, etc., Donate old clothes to the needy, instead of throwing them away.
- If you are sure the item is not usable in its present form, can it be recycled? Paper, plastics, glass, metal can all be recycled.

- Segregate your waste into wet and dry garbage. Wet garbage includes most kitchen wastes. Most of this can be used for composting. Most dry garbage is recyclable. The amount of dry waste generated in your household is an indicator of how well you are following the 3Rs principle. A lot of dry waste means you should go back to the 'Reduce and Reuse' principles and try to follow them better.
- Avoid the use of non-biodegradable materials such as Styrofoam and certain types of plastics. Although most plastics are recyclable, recycling still takes up energy, which is another precious resource not to be wasted. If thrown away as waste, Styrofoam and plastics can take hundreds of years to decompose.
- Do not litter or throw garbage in public places. Garbage and litter is a visual contaminant and can cause diseases and health problems. Proper disposal of garbage is an important part of waste management.
- Be a conscious consumer and do not buy products that are over packaged. Try choosing products that are made from recycled material or are organically grown.

Enforcement of Environmental Legislation

Environmental legislation is evolved to protect our environment as a whole, our health, and the earth's resources. The presence of a legislation to protect air, water, soil, etc., does not necessarily mean that the problem is addressed. Once legislation is made at the global, national or state level, it has to be implemented. For a successful environmental legislation to be implemented, there has to be an effective agency to collect relevant data, process it and pass it on to a law enforcement agency. If the law or rule is broken by an individual or institution, this has to be punished through the legal process. Information to law enforcement officials must also come from concerned individuals. In most situations, if no cognizance is given, the interested concerned individual must file a Public Interest Litigation (PIL) for the protection of the environment. There are several NGOs in the country such as WWF-I, BEAG and the BNHS which take these matters to court in the interest of conservation. Anyone can request them to help in such matters. There are also legal experts who have successfully fought cases in the courts to support environmental causes. A related issue is the fact that there are several irregular practices for which a bribe to an unscrupulous official is used to cover up an offence. Thus, the general public must act as a watchdog not only to inform concerned authorities, but also to see that actions are taken against offenders.

Environment Impact Assessment (EIA)

For all development projects, whether government or private, the MoEF requires an impact assessment done by a competent organization. The EIA must look into physical, biological and social parameters. EIAs are expected to indicate what the likely impacts could be if the project is passed. The MoEF has identified a large number of projects that need clearance on environmental grounds. The EIA must define what impact it would have on water, soil and air. It also requires that a list of flora and fauna identified in the region is documented and to specify if there are any endangered species whose habitat or life could be adversely affected. Most development projects such as industries, roads, railways and dams may also affect the lives of local people. This must also be addressed in the EIA. There are 30 different industries listed by MoEF that require a clearance before they are set up. Impacts created by each type of industry differ and the proposed sites also vary in their sensitivity to impacts. Some areas are more fragile than others. Some have unique ecosystems. Others are the habitats of wildlife and some may be the home of endangered species of plants or animals. All these aspects require evaluation before a development project or an industry site is cleared. New projects are called 'green field projects' where no development has been done. Projects that already exist but require expansion must also apply for clearance. These are called 'brown field projects'. After the Environmental Protection Act of 1986 was passed, an EIA to get an environmental clearance for a project became mandatory. Project proponents are expected to select a competent agency to undertake an EIA. Projects can be classified into those with a mild impact, a moderate impact or a serious impact. Some may have temporary major impacts, during the construction phase, which could later become less damaging, or be mitigated by a variety of measures. In other situations the impact may continue and even increase, for example, where toxic solid waste will be constantly generated. Some projects could thus cause temporary reversible damage while others can have irreversible or even permanent impacts. To get an environmental clearance the proposer of the project is expected to apply to the State Pollution Control Board (PCB). The PCB checks and confirms that the EIA can be initiated. The agency that does the assessment submits a report to the proposer. This may take several months. The report is then forwarded to the MoEF, which is the impact assessment authority. After 1997, the MoEF has stipulated that a public hearing should be done at the local level. The PCB puts an advertisement about the hearing in the local vernacular press. An Environmental Impact Statement, which is an executive summary of the EIA is kept for the public to read. The venue and time of the public hearing is declared. Once the hearing is held and opinions have been expressed, both for and against the project, the minutes of the meeting are sent to the MoEF. Though this is done, it is evident that the voices of project-affected people are still not heard. In some cases NGOs have taken up the cause of local people. Until educational levels and environmental awareness become a part of public thinking and are objectively based on the facts of the case, these hearings will remain an inadequate tool to control possible impacts of new development projects. Experience shows that a large number of EIAs are inadequately researched and frequently biased as they are funded by the proposer of the project. While most EIAs are adequate for studies on the possibilities of air, water and soil pollution, they generally deal inadequately with issues such as preservation of biodiversity and the social issues that may arise from future environmental impacts.

Biodiversity concerns frequently are sketchily considered and mostly consist of a listing of species without population assessments, or census figures of wildlife, or a study of the affects on the ecosystem as a whole. Changes in land use patterns affect whole communities of living organisms. This is rarely taken into account, as such issues are difficult to assess in quantifiable terms. Issues related to equity of resources that are inevitably altered by development related projects are also not fully addressed. These cryptic concerns must be dealt with more seriously in environmental assessments and the public at large should know and appreciate these inadequacies. It is not sufficient to say that an EIA has been done. It is the quality and sincerity of the EIA that is of importance. An EIA is not intended to stop all types of development. The site of an industry can be selected carefully and if it is likely to damage a fragile area an alternate less sensitive area must be selected. In some cases it is essential to drop projects altogether if the anticipated impacts are likely to be very severe. In other cases it is necessary for the project to counter-balance its effects by mitigating the ill-effects on the environment. This means compensating for the environmental damage by afforestation or creating a protected area in the neighbourhood at the cost of the project. Rehabilitation and resettlement of project-affected people is a key concern which should be given adequate funds and done after consent is clearly obtained from the people living in the area. In most cases it is advisable to avoid resettlement altogether. If an area's vegetation is being affected project costs must include the cost of compensatory afforestation and other protective measures.

Public Awareness

Environmental sensitivity in our country can grow only through a major public awareness campaign. This has several tools. The electronic media, the press, school and college education and adult education are all essentially complementary to each other. Green movements can grow out of small local initiatives to become major players in advocating environmental protection to the government. Policy makers will only work towards environmental preservation if there is a sufficiently large bank of voters that insist on protecting the environment. Orientating the media to project pro-environmental issues is an important aspect. Several advertising campaigns frequently have messages that are negative to environmental preservation.

Environmental Education

There are many definitions of environmental education. The one that has been adopted by the government in the Environment 2010 Strategy and the National Strategy for Environmental Education is: A multidisciplinary approach to learning that develops the knowledge, awareness, attitudes, values and skills that will enable individuals and the community to contribute towards maintaining and improving the quality of the environment. The common elements of this and other definitions for environmental education include:

- the influence of environmental education on values, attitudes and behaviour;
- the multidisciplinary nature of environmental education and the emphasis on linkages between the biophysical environment, and socio-economic and political activities;
- the contribution of environmental education to protecting and managing the environment; and
- the range of learning activities encompassed by environmental education that include formal and non-formal education;

Environmental education is therefore not just about classroom learning but about all learning activities intended to inform people about the environment and its management. The term 'environmental education' embraces a wide range of learning activities that inform people about resource management and the environmental effects of resource use. Thus, there can be many providers of environmental education. Within the formal education sector, these providers can include primary, intermediate and secondary schools, the colleges of education, the universities, polytechnics and other tertiary institutions.

Outside the formal education sector there is, once again, a wide variety of providers. These include government departments undertaking specific programmes or less formal activities to promote the implementation of policy. Also included are local authorities, both regional and territorial, using environmental education as a tool to achieve objectives and policies outlined in planning documents. Providers also extend well beyond central and local government. Research institutes, environmental non-government organizations, private institutions, landcare and other community groups, and the environmental consulting industry are amongst those that undertake activities that may be seen as environmental education. In regard to governmental responsibilities, an effective policy framework for protecting and managing the environment requires an integrated approach that includes sound legislation, good knowledge about the environment, the communication of information, and responsible behaviour by individuals, resource users, and communities. Within this framework, environmental education contributes to the communication of information and to the development of understanding, skills, attitudes and values that influence the behaviour of individuals, resource user groups, and the community.

The National Strategy for Environmental Education advocates a three-fold approach to effective environmental education, namely, education about the environment, in the environment, and for the environment. Education about the environment is concerned with providing information on environmental phenomena – the functioning of natural systems and the interactions between humans and their environment. This informative approach is concerned with developing awareness, knowledge and understanding of the environment.

Education in the environment is experiential; it is activity based, located outside the classroom setting, and encourages knowledge through contact with the environment, for example, using field studies and other outdoor activities for learning and skill development.

Education for the environment has environmental improvement as its goal. This involves specific actions to improve environmental quality. It can develop a sense of responsibility and empower people to feel that they can make a difference by being actively involved in measures to improve environmental outcomes.

The above approach aims to assist people to:

- increase awareness of sustainable resource management and environmental stewardship;
- acquire a greater understanding and knowledge about the environment and human interaction with it;
- foster attitudes and values of care for the environment and the natural and physical resources of the biosphere;
- obtain the skills required for effective resource management and increased environmental protection;
- participate in decision-making and actions to improve the sustainability of resource use and the quality of the environment.

The linkages between these various aspects of environmental education are shown in Figure 13.1. The underlying assumption to the above approach is that environmental education must do more than simply increasing knowledge about the environment. It must also help develop environmentally responsible attitudes and values, a commitment to work for change, and a wide range of skills for successfully addressing resource management and environmental issues. The need for collaborative planning and shared decision-making with the community is essential for achieving these. The use of environmental education to promote sustainable resource management and improved environmental protection is also based on the realization that, not only is it more effective in the long term to develop awareness and understanding, promote changes in attitudes and values, and provide skills for participation in resource management processes and decision-making, but that this approach, in fact, reflects community aspirations. For example, the Wellington Regional Council (1998) has undertaken informal surveys of farmers, business people, individuals and community groups within its region and found that "all groups rank environmental education rather than regulation as the preferred role for the Council as an environmental management authority".

Environmental education, although it may impose costs initially, has the potential to reduce not only regulation *per se,* but also regulatory and enforcement costs over time by promoting long-term changes in behaviour in relation to resource use, development and protection. Spending on environmental education today, therefore, should be viewed as an investment in the future, rather than a cost imposed upon the present. The following principles merit particular attention when developing, implementing and evaluating environmental education:

- *Comprehensiveness:* A diverse and comprehensive range of educational activities should be used to achieve the desired outcomes. This involves an interdisciplinary approach to environmental education.
- *Collaboration:* Environmental education should encompass considerations of all viewpoints and should involve other organizations, the community and tangata whenua in implementing the programme.
- *Coordination:* Coordination with other agencies providing environmental education is also important.

- *Local and relevant:* Environmental education should be centred on 'real world' resource management problems and issues that are local and relevant to people's everyday lives.
- *Action focus:* Emphasis should be given to developing skills and providing opportunities for practical experience. Such actions tend to be of more lasting impression to participants than simply passive initiatives.
- *Benefits and solutions:* Environmental education should focus on benefits and solutions, rather than simply problems. Often environmental problems appear intractable. This engenders a sense of powerlessness. Accentuating benefits and solutions, on the other hand, can encourage feelings of empowerment.

Figure 13.1 Environmental Education and Its Entities

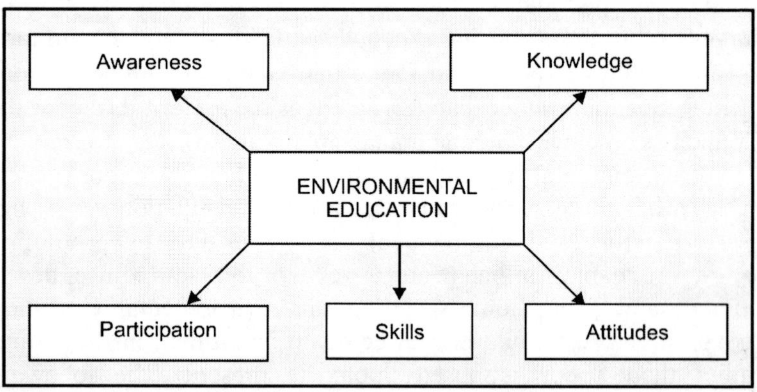

Commitment to Environmental and Social Issues

There is an increasing awareness and sensitivity the world over to the environmental implications of development. This has led to the strengthening of environmental legislation and enforcement as well as increased public pressure on projects and promoters to demonstrate the environmental soundness of projects. The emerging trends in India, such as tightening environmental laws and enforcement, increased public pressures and litigation, indicate that environmental risks have already started assuming significance.

Over the past years, the business case for environmental and social management at financial institutions has been made.

Environmental and social (including labour) risks with clients translate into credit and reputation risk for financial institutions financing such clients. However, in general, businesses served by micro-finance are not associated with the same scale and types of environmental and social risks as larger corporate clients and projects served by other financial institutions. Moreover and naturally, the focus in micro-finance lies on the development of positive social impact (and the ways of measuring that).

Nevertheless, when considering the broad spectrum of micro-entrepreneurs, specific environmental and social risks may not be overlooked. Child labour, groundwater pollution by use of agro-chemicals, soil pollution by an automotive repair workshop are some examples. The adverse environmental and social impacts of one micro-entrepreneur may be small, but since the number of clients served is usually large, there is a significant cumulative effect.

Table 13.3 Loan Cycle, Existing Format and E&S Aspects

Loan Cycle	*Existing Formats*	*E&S Aspects*
Application	Loan application	Exclusion list
Appraisal	Loan appraisal format	Activity assessment tool/Fact sheets
Contracting and disbursement	Loan contract	Contract clauses
Recovery and monitoring	MIS format	Monitoring specifics
Reporting	Reports	Reporting formats

Application: At the time of application, purpose of loan is usually asked. The exclusion list should be used at this point to reject applications whose stated purpose falls within activities in the exclusion list.

Appraisal: In addition to existing items in the appraisal format, MFIs should include information on possible environmental and social risks associated with the said activity. The activity assessment tool and sector fact sheets will be used here.

There are three possible outcomes of social and environmental appraisal depending on environmental and social risk level and size of loan. The MFI could choose to Raise client's awareness about social and environmental aspects:

- educate the client regarding social and environmental improvements;
- include specific clauses in the loan contract to mitigate social and environmental risks; and
- loan contracting and disbursement: In general, standard clauses like proper disposal of waste, hygienic working and living conditions, etc., could be introduced in the loan agreement. If a specific risk is identified at the time of appraisal, a special clause could be included in the loan agreement to mitigate the risk.

Recovery and monitoring: For effective monitoring social and environmental aspects pertaining to a loan will need to be recorded in the MIS of the organization. Some examples are:

- Whether an E&S appraisal has been performed.
- Key E&S aspects.
- Additional clauses to contract, if any.
- Any improvements made.

This information could be used for analysis of improvements and reporting.

Reporting: Reporting formats will be developed. Information regarding the following would be provided:

- Total loans disbursed for the year.
- Number of applications.
- Number of rejections based on exclusion list.
- Number of E&S interventions (awareness raising, contract clause).
- Common risks.
- Most risky sector.

Issues to be Addressed

While evaluating the environmental and social risks associated with potential investees, consider the general environmental and social issues that are common to all sectors that the investee MFI may be financing. Environmental factors that would be considered are:

Table 13.4 Major Environmental Factors and Objectives

Factor	*Objective*
Air Management	Ensure potential air pollutants are contained and activities do not impact the natural environment.
Noise Management	Ensure noise/vibration levels meet statutory requirements and acceptable standards.
Water Management	Maintain or improve quality of surface water. Maintain or improve quality of ground water.
Wastewater Management	Protect aquatic ecosystems, reuse treated wastewater on site.
Waste Management	Ensure wastes are disposed of in a proper manner and should not pollute the immediate environment.
Hazardous Materials Management	Ensure chemicals are stored and disposed off carefully. Ensure pest control chemicals are used safely.
Contaminated Land	Ensure land is not contaminated through financed activities.

Both the environmental and social issues should be taken into account altogether. These two issues are one and cannot be separated. Due to tremendous growth in population and thus intensive man induced activities, environmental degradation was severe and consequently its impact on society was considerable. As a result of this, environmental and social issues got momentum in due course of time particularly after 1970s when major environmental protection movements and resolutions started in India and abroad. There is thus urgent need to solve the two issues – environmental and social – through sustainable development of natural resources and control over carbon emissions.

References

Economic Survey 2005–2006.

Human Development Report, 1998, pp. 67, 79.

Report of Ministry of Environment and Forest, Government of India.

CHAPTER 14

Conservation Movements

The conservation movement, also known as nature conservation, is a movement that aims to protect natural resources including plant and animal species as well as their habitats for the future. The early conservation movement included fisheries and wildlife management, water, soil conservation and sustainable forestry. The contemporary conservation movement has broadened itself from the early movement's emphasis on use of sustainable yield of natural resources and preservation of wilderness areas to include preservation of biodiversity. Some environment experts are of the view that the conservation movement is part of the broader and more far-reaching environmental movement, while others argue that they differ both in ideology and practice. In the United States, conservation is mainly seen as differing from environmentalism in that it aims to preserve natural resources expressly for their continued sustainable use by humans. In other parts of the world conservation is used more broadly to include the setting aside of natural areas and the active protection of wildlife for their inherent value, as much as for any value they may have for humans.

Jones (1991) argues that from an economic perspective the western nations have been no more destructive to natural resources than any other civilization. He rejects the suggestion that Christianity, by destroying animism, facilitated the ruination of nature in the West, stating he finds no empirical evidence that any culture was or is less exploitive of the natural world than Christianity. He notes that Eastern agricultural history has numerous examples of massive deforestation, erosion, silting of rivers, and infestation with water-borne parasites. He points to large-scale animal extinction and wasteful agricultural practices by North American Indians before 1492. Jones says that economic growth in the West did result in a higher level of resource use, but finds no evidence to support the view that such resource exploitation was a product of religion, culture, or geography.

History of Conservation Ideas

The nascent conservation movement slowly developed in the 19th century, starting first in the scientific forestry methods pioneered by Prussia and France in the 17th and 18th centuries. While continental Europe created the scientific methods later used in conservationist efforts, British India and the United States are credited with starting the conservation movement. Foresters in India, often German, managed forests using early climate change theories that Alexander von Humboldt developed in the mid-19th century, applied fire protection, and tried to keep the 'household' of nature. This was an early ecological idea, in order to preserve the growth of delicate teak trees. The same German mandarins who headed the Forest Service of India, such as Dietrich Brandis and Berthold Ribbentrop, travelled back to Europe and taught at forestry schools in England. They brought with them the legislative and scientific knowledge of conservationism in British India back to Europe, where they disseminated widely to people like Gifford Pinchot, who in turn helped bring European and British Indian methods to the United States.

The concept of environmentalism was postulated by John Muir (1838–1914). Muir's passion for nature made him the most influential American environmentalist. Muir preached that nature was sacred and humans are intruders who should look but not develop. He founded the Sierra Club and remains an icon of the environmentalist movement. He was primarily responsible for defining the environmentalist position, in the debate between conservation and environmentalism. It allowed for limited tourism, but opposed automobiles in national parks. It strenuously opposed timber cutting on most public lands, and vehemently denounced the dams that President Theodore Roosevelt supported for water supplies, electricity and flood control. Especially controversial was the Hetch Hetchy dam in Yosemite National Park, which Roosevelt approved, and which supplies the water supply of San Francisco.

Sivaramakrishnan (2009) explores the boundaries between wildness and civility in Indian society as well as connection of ideas of nature to different aspects of social life, especially labour, aesthetics, politics, commerce and agriculture. These interconnected historical processes inform environmental history in India. At present, forest history is the area of environmental history in which the most important scholarly debate is underway in India, with special interest in questions of water, air, industry and climate change.

Vaccaro and Beltran (2009) examine the Pyrenees mountains as an environmental reservoir, where most of the population has emigrated and the state has taken over much of the mountainous territory to implement conservation policies. One result is the return of fauna, via reintroduction or natural regeneration, especially bears, wolves, beavers, river otters, marmots, mouflon, feral goats and deer. The work and living space of the mountain communities has fallen under the jurisdiction of external institutions and constituencies that value conservation and eco-tourism above local subsistence. This has led to controversy among scientists and residents. Local herders see wild animals as unregulated public property subsidized by the work of the local people. Farmers complain that their fields are invaded on a daily basis by animals they cannot kill because of their protected status. Ranchers, under extremely strict sanitation regulations, see their cattle and sheep coming into biological contact with these unchecked wild populations.

In Russia, the regime of Joseph Stalin (1924–53) concentrated on large-scale industrialization, and has earned a historical reputation for paying little heed to the human and environmental costs of such rapid transformations. One exception was the logging industry, in which conflicting needs and ideologies enabled the preservation of significant tracts of forest. Galbreath and Auers

(2009) examine the history of environmental politics in Latvia, especially the formation of the Latvian Green Party, known as Zala Partija as part of the Green/Farmers' Union or Zalo un Zemnieku Savieniba (ZZP). The depoliticization of environmentalism emerged from the nationalist, corporate and environmental elements of the ZZP. The three aspects of the environmentalism here are represented by the green, brown and black colours. Green represents ecological preservation and reversal of industrial side effects, brown stands for civic and ethnic nationalism and black represents the oil and gas pipeline industry and its influence on the 'green' agenda in Latvia.

Both conservationism and environmentalism appeared in political debates in the US during the progressive era in the early 20th century. There were three main positions. The laissez-faire position held that owners of private property, including lumber and mining companies, should be allowed to do anything they wished for their property. The conservationists, led by President Roosevelt and his close ally Gifford Pinchot, said that the laissez-faire approach was too wasteful and inefficient. In any case, they noted, most of the natural resources in the western states were already owned by the federal government. The best course of action, they argued, was a long-term plan devised by national experts to maximize the economic benefits of natural resources.

Roosevelt put the conservationism issue high on the national agenda. He worked with all the major figures of the movement, especially his chief advisor on the matter, Gifford Pinchot. Roosevelt was deeply committed to conserving natural resources, and is considered to be the nation's first President who has championed the cause of conservation of environment. He encouraged the Newlands Reclamation Act of 1902 to promote federal construction of dams to irrigate small farms. He also set aside more federal land for national parks and nature preserves than all of his predecessors combined. He established the United States Forest Service, signed into law the creation of five National Parks, and also signed the 1906 Antiquities Act, under which he proclaimed 18 new US National Monuments. He also established the first 51 Bird Reserves, four Game Preserves, and 150 National Forests, including Shoshone National Forest, the nation's first. The area of the United States that he placed under public protection totals approximately 230 million acres (930,000 km^2).

In 1903 Roosevelt toured the Yosemite Valley with John Muir, who had a very different view of conservation, and tried to minimize commercial use of water resources and forests. Working through the Sierra Club he founded, Muir succeeded in 1905 in having Congress transfer the Mariposa Grove and Yosemite Valley to the National Park Service. While Muir wanted nature preserved for the sake of pure beauty, Roosevelt subscribed to Pinchot's formulation, "to make the forest produce the largest amount of whatever crop or service will be most useful, and keep on producing it for generation after generation of men and trees". In the 1930s, the dominant view was the conservationism of Theodore Roosevelt, endorsed by Democrat Franklin D. Roosevelt that led to the building of many large-scale dams and water projects, as well as the expansion of the National Forest System to buy out sub-marginal farms.

Environmental issues re-emerged on the US national agenda in 1970, with Republican President Richard Nixon playing a major role, especially with his creation of the Environmental Protection Agency. The debates over the public lands and environmental politics played a supporting role in the decline of liberalism and the rise of modern conservatism. Although Americans consistently rank environmental issues as "important", polling data indicates that in the voting booth voters rank the environmental issues low relative to other political concerns. Deforestation and overpopulation are issues affecting all regions of the world. The consequent destruction of wildlife habitat has prompted the creation of conservation groups in other countries, some founded by local hunters, who have witnessed declining wildlife populations first hand. Also, it was highly important for the conservation movement to solve problems of living conditions in the cities and the overcrowding of such places.

The idea of 'incentive conservation' is a modern one but its practice has clearly defended some of the sub-Arctic wildernesses and the wildlife in those regions for thousands of years, especially by indigenous peoples such as the Evenk, Yakut, Sami, Inuit and Cree. The fur trade and hunting by these peoples have preserved these regions for thousands of years. Ironically, the pressure now upon them comes from non-renewable resources such as oil, sometimes to make synthetic clothing which is advocated as a humane substitute for fur. Similarly, in the case of the beaver, hunting and fur trade were thought to bring about the animal's demise, when in fact they were an integral part of its conservation. For many years children's books stated and still do that the decline in the beaver population was due to the fur trade. In reality, however, the decline in beaver numbers was because of habitat destruction and deforestation, as well as its continued persecution as a pest (it causes flooding). In Cree lands, however, where the population valued the animal for meat and fur, it continued to thrive. The Inuit defend their relationship with the seal in response to outside critics.

In order to discourage illegal South African hunting parties and ensure future local use and sustainability, indigenous hunters in Botswana began lobbying for and implementing conservation practices in the 1960s. The Fauna Preservation Society of Ngamiland (FPS) was formed in 1962 by: Robert Kay and June Kay – the environmentalist couple working in conjunction with the Batawana tribes to preserve wildlife habitat. The FPS promotes habitat conservation and provides local education for preservation of wildlife. Conservation initiatives were met with strong opposition from the Botswana government because of the monies tied to big-game hunting. In 1963, BaTawanga chiefs and tribal hunters/adventurers in conjunction with the FPS founded Moremi National Park and Wildlife Refuge – the first area to be set aside by tribal people rather than governmental forces. Moremi National Park is home to a variety of wildlife, including lions, giraffes, elephants, buffalo, zebra, cheetahs and antelope, and covers an area of 3,000 square kilometres. Most of the groups involved in establishing this protected land were associated with hunting and were motivated by their personal observations of declining wildlife and habitat (Encarta Encyclopedia, 2005).

Leaders of Conservation Movements in India

Anna Hazare

Anna Hazare is an environmental and social activist. He has been working for social and environmental welfare of the people since three decades. He received many awards for his contribution towards conserving the nature. He is a leading activist and the hero of poor rural people. He adopted Gandhian philosophy and was sentenced to imprisonment several times for his peaceful agitation. Recently, he led an agitation – for formation of Jan Lokpal Bill. This was supported nationwide and after 11 days of his fast unto the death, the Government of India have to agree on his demands.

Chandi Prasad Bhatt

Chandi Prasad Bhatt is also a Gandhian environmentalist and social activist, founded *Dasholi Gram Swarajya Sangh* (DGSS) in Gopeshwar in 1964 that became the mother organization to the Chipko movement. He has been awarded the Magsaysay Award for Community Leadership in 1982, followed by the Padma Bhusan in 2005. He is known for his work on subaltern social ecology and considered one of India's first modern environmentalists. Bhatt felt deeply concerned over the plight of the mountainous people as a whole, and he often walked through the mountainous to talk to the villagers about their problems. Among the most important, of course, were the shortages in farmland and jobs. But added to these were oppressive government policies concerning the forests. The villagers depended on the forests for firewood, fodder for their cattle, and wood for their houses and farm tools. But, the government restricted huge areas of forests from their use, and then auctioned off the trees to lumber companies and industries from the plains – a practice inherited with little change from the British colonialists. Because of these restrictions and an ever-growing population, the mountain women found themselves walking hours each day just to gather firewood and fodder. Curtailment of the villagers' legitimate rights to trees and forest products in favour of outside commercial interests enabled Bhatt, in 1973, to mobilize the forest-wise society members and villagers into the collective Chipko Andolan (Hug the Trees Movement) to force revision of forest policies dating from 1917. In the process he became knowledgeable and productive in helping ensure his people's hard-won living. In 2003, he was appointed a member of the National Forest Commission, which reviewed all existing policies and legal frameworks relating to forest management, and submitted its report to the government in 2005.

Chewang Norphel

Chewang Norphel is a pioneer activist and a social worker who spend his life in the cold desert of Ladakh for the welfare of local farming community. He developed irrigation system in this region, which helped the farmers to enhance their crop production. Falling in the rain-shadow area of the Himalayas, Ladakh is a cold, mountainous desert. During winter, temperature drops below the –30°C mark. Annual average rainfall is 50 mm. The only source is glacier water from the mountains. They melt in summer, releasing the little water that the people of Ladakh have to manage with the utmost care. It is stored for use throughout the year. It's small wonder that Norphel realized the importance of water. Coming from a middle class family of Leh, Norphel went to Amar Singh College in Srinagar as a student of science. He completed a diploma course in civil engineering from Lucknow in 1960. In June 1960, he joined the rural development department of Jammu and Kashmir in Ladakh as a civil engineer.

In 1996, one year after he had retired, Norphel joined the Leh Nutrition Project, a non-governmental organization, as project manager for watershed development. "Watershed development is the only solution for Ladakh's rural economy. Otherwise, you will have rural folk flocking the city in search of jobs. And there are not many to go around," he indicates. Norphel proposes artificial glaciers as a substitute for dams, which to him are environmental and social problems, besides being a financial burden. Norphel has brought glaciers closer to the villages and water comes in early as a bonus for the farmers.

Mahesh Chandra Mehta

Mahesh Chandra Mehta is a public interest attorney. He was awarded the Goldman Environmental Prize in 1996 for his continuous fights in Indian courts against pollution-causing industries. He also received the Magsaysay Award for Public Service in 1997.

Medha Patkar

Born in Mumbai Medha Patkar was raised by politically and socially active parents. Her father actively participated in the independence movement. Her mother was a member of Swadar, an organization set up to help and assist women suffering difficult circumstances arising out of financial, educational and health-related problems. Her parents' activism played a major role in shaping her philosophical views. She did her M.A. in Social Work from Tata Institute of Social Sciences, Mumbai and then she worked for seven years with voluntary organizations.

Veer Bhadra Mishra

Veer Bhadra Mishra is the founding president of the Sankat Mochan Foundation (SMF). He is a former professor of Hydraulic Engineering and former Head of the Civil Engineering Department at the ITBHU, Varanasi. He is also the Mahant (high priest) of the Sankat Mochan Hanuman Temple, Varanasi founded by poet-saint Goswami Tulsidas. Mishra was recognized on the United Nations Environment Programmes (UNEP) "Global 500 Roll of Honour" in 1992, and was a TIME magazine's "Hero of the Planet" recipient in 1999 for his work related to cleaning of Ganga through the SMF. His education has helped him understand threats to the Ganges and since 1982 he has struggled to open the eyes of bureaucrats and the public. Supported in part by aid from the various individuals and some government agencies of the United States and Sweden, Mishra juggles his roles as priest and activist. Motivated most of all by "respect and love for the river," Mishra, working with William Oswald, an engineering professor emeritus at the University of California, Berkeley, proposed what is called an Advanced Integrated Wastewater Oxidation Pond System (AIWPS). It would store sewage for 45 days, using bacteria and algae to eliminate waste and purify the water.

Sunderlal Bahuguna

Sunderlal Bahuguna is a noted environmentalist, Chipko movement leader and a follower of Mahatma Gandhi's philosophy of non-violence and *satyagraha*. For years he has been fighting for the preservation of forests in the Himalayas, first as a member of the Chipko movement in 1970s, and later spearheaded the movement against the construction of Tehri dam starting in 1980s, to early 2004. He was one of the early environmentalists of India, who with people associated with the Chipko movement later started taking up environmental issues, like large dams, mining and deforestation, across the country. He was awarded the Padma Vibhushan, India's second highest civilian honour, on January 26, 2009.

Arvind Kumar Gupta

Arvind Kumar Gupta, after graduating from IIT, Kanpur, in 1975, joined TELCO, Pune and worked there for a few years. In 1978, he took one year's study leave to work with the grass roots village science teaching programme for children in the tribal district of Hoshangabad, Madhya Pradesh. There, he developed many useful low-cost teaching/science teaching aids using locally available materials. The possibilities of using ordinary things for doing science and recycling modern junk into joyous products appealed immensely to children. Gupta's first book "Matchstick Models and Other Science Experiments" was translated into 13 Indian languages by various popular science groups and sold more than half a million copies. As a UNESCO consultant on science education, he has been invited to share his experiences in science teaching with teachers of several developing countries. He has been actively associated with the Bombay Natural History Society, Conservation Society of Delhi, Spastic Society of North India and the Bharat Gyan Vigyan Samiti. He is also an advisor to the National Book Trust on popular science books. He has received several awards for his outstanding contributions in the field of science education. These include Eklavya Award (1982), the National Award for Science Popularisation amongst Children (1988), Hari Bhau Mote Award of the Marathi Vigyan Parishad (1988), a special award given by the National Association for the Blind for designing teaching aids for pre-school blind children (1991), Granthali award for his book *Khel* (1992), Ruchi Ram Sahni Award for science popularization (1993) and the Hari Om Ashram Award by the UGC (1995).

Major Conservation Movements in India

Chipko Movement

The Chipko movement is a social-ecological movement that practised the Gandhian methods of *satyagraha* and non-violent resistance, through the act of hugging trees to protect them from being felled. The movement started in the early 1970s in the Garhwal Himalaya of Uttarakhand, with growing awareness towards rapid deforestation. The landmark event in this struggle took place on March 26, 1974, when a group of peasant women in Reni village, Hemwalghati, in Chamoli district, Uttarakhand, acted to prevent the cutting of trees and reclaim their traditional forest rights that were threatened by the contractor system of the state forest department and inspired hundreds of such grass roots level actions throughout the region. Gaura Devi of Reni village led this movement and later on Sunder Lal Bahuguna and Chandi Prasad Bhatt spread this throughout the region. By the 1980s, the movement spread throughout India, and led to formulation of people-sensitive forest policies and stopping of open felling of trees in regions as far reaching as Vindhyas and the Western Ghats. The first recorded event of Chipko however took place in village Khejarli, Jodhpur district, in 1730 AD, when 363 Bishnois, led by Amrita Devi, sacrificed their lives while protecting green Khejri trees, considered sacred by the community, by hugging them.

Clean Ganga Campaign

The campaign has been launched to keep the river Ganga clean and pollution-free. However, it could not succeed much in achieving its goals and has become merely an awareness campaign. It was the Sankat Mochan Foundation (SMF) which took the responsibility to keep the river clean. SMF was established in the year 1982 as the non-profit, non-political organization under the Societies Act of Government of India. Its major objective is to make the river clean by the fast deteriorating environmental conditions, and promote education and health care programmes for the unprivileged class. The SMF also carries the age-old cultural traditions of the Varanasi. It runs the campaign 'Swachha Ganga' (Clean Ganges) programme.

The Government of India has launched the Ganga Action Plan in the year 1986. It did cost Rs 5 billion, but did not have any successful result. The government has then put the scheme of Ganga Clean under the Ganga Action Plan, which is successful, but the real result of the scheme is narrated by the scientific data. The higher the biochemical oxygen demand, the higher will be the pollution. The SMF AIWPS Solution was invented by the Professor William J. Oswald and his co workers at the University of California. In the year 2004 'Clean Ganga – Clean Ghat' is the exhibition project that is implemented with the support of the British High Commission, New Delhi under the SIDA project 'Clean Ganga Conclave' at Tulsighat under the CSPP Clean Ganga Day that is organized at New Delhi.

Himalaya Bachoa Andolan

The environmental activists throughout the Himalayan regions have been working to save Himalaya through launching various programmes. These programmes include afforestation, and agitating against construction of big hydropower projects. Agitation against the Tehri dam, led by the environmentalist Sunder Lal Bahuguna, is an example of this movement.

Narmada Bachao Andolan

Veteran social activist Medha Patekar led the Narmada Bachao Andolan during the 1990s against a big dam on the Narmada river that affected thousands of people coming under submergence area.

Climate Change Conventions

The issues of climate change were first discussed in the Rio de Janeiro Earth Summit in 1992. Two concepts were postulated in this summit. While the first pertained to different responsibilities of the nations, in the second, the world countries were advised to work together for reduction in carbon emissions. Likewise, while the industrial countries have to do this as a responsibility, the developing countries should try to cut it keeping their development activities in view. In addition, developed countries will assist developing countries financially for this purpose. Thereafter, a climate summit was held in Kyoto in 1997, which is known as Kyoto Protocol. A target was fixed under UN Framework for Convention of Climate Change (UNFCCC). To fill up this target, developed countries were advised to reduce carbon emission upto the level of 1990. The first phase was fixed between 2008 and 2012. All countries accepted this protocol but the United States of America and Australia did not sign.

Inter-Governmental Panel for Climate Change (IPCC) submitted its report to the UN regarding global warming and climate change in 2007. Rajendra Pachori, who heads this panel, received Nobel Peace Prize of 2008 sharing this with El Gore of the USA. IPCC report warned the world community that the crisis of climate change is being deepened. The report indicated that during the last 100 years, the earth temperature increased 0.4° to 0.6°C. It further stated that after industrial revolution 1°C temperature has increased. The report indicated, if the trends of present global warming are continued, the earth's temperature will increase 1° to 4°C by 2100. It has been too late to control it completely but there is still a time to reduce it upto a half level, the report says. Discussions on IPCC report and climate change are continued at various foras.

In 2008, the Obama administration in the US changed the main framework of economic forum which was supported by the developed countries of European Union. The proposal stated that the developing countries should also cut carbon emissions. Dr Man Mohan Singh, Prime Minister of India, in response to this proposal, stated that per person carbon emission in India is not higher than the developed countries. Recently, both the USA and European Union have warned India to reduce the carbon emission at a higher rate than targeted. Major summits and conferences are discussed in the following paragraphs:

Earth Summit

In 1992, the UN Conference on Environment and Development was held in Rio de Janeiro, Brazil. Popularly known as the Earth Summit, this meeting was the largest gathering of world leaders in history. The conference produced two major treaties. The first was an agreement for nations to voluntarily reduce emission of gases leading to global warming, and the second was a pact on biodiversity requiring countries to develop plans to protect endangered species and habitats. At the insistence of the United States, however, the final version of the global warming treaty was dramatically scaled back. The United States was also one of the very few countries that refused to sign the biodiversity treaty. It objected to a part of the treaty that specified that money to come from the use of natural resources from protected ecosystems, such as rainforests, should be shared equally between the source country and the corporation or institution removing the materials.

The 1992 agreement on global warming limited each industrialized nation to emissions in the year 2000 that were equal to or below 1990 emissions. However, these limits were voluntary and with no enforcement provisions included in the agreement. It became clear by 1997 that these goals would never be met. At a follow-up conference in Kyoto, Japan, representatives from 160 countries signed the Kyoto Protocol. This agreement called for industrialized nations to reduce emissions to an average of about 5 per cent below 1990 emission levels and to reach this goal between the years 2008 and 2012. For this accord to become international law, however, it had to be ratified by at least 55 countries.

Montreal Protocol

After the Earth's Summit of 1992, Montreal Protocol was signed in 1995. Depletion of ozone layer and its reduction strategy was the major issue in this meeting. In response to regulations enforced by the Montreal Protocol on substances that deplete the ozone layer and its amendments, the atmospheric concentrations of many of these gases are either increasing more slowly or decreasing.

World Summit on Sustainable Development

In 2002, delegates from nearly 200 countries convened at the World Summit on Sustainable Development in Johannesburg, South Africa, to establish new sustainable development goals for the 21st century. They also negotiated to strengthen commitments from the governments of developed nations to provide aid for sustainable development. Among the outcomes, the summit created an action plan that called on nations to reduce by half the proportion of people who lack sanitation by 2015, to minimize health and environmental problems caused by chemical pollution by 2020, and to reduce significantly the number of endangered species by 2010.

Kyoto Protocol

At the December 1997 international summit in Kyoto, Japan, world leaders for the first time agreed to specific targets and tentative time tables for reducing greenhouse gas emissions. Scientists have long warned that such emissions could cause significant global warming in the decades ahead. Kyoto summit was named 'The Third Conference of Parties to the Framework Convention on Climate Change'. The meeting had its origins in the UN's 1992 Earth Summit in Rio de Janeiro, Brazil. Thousands of diplomats, non-governmental organization (NGO) observers, journalists and political leaders from more than 150 nations attended the summit. Prime Minister Ryutaro Hashimoto of Japan and Vice President Al Gore of the United States delivered formal statements in this meeting. In the end, delegates negotiated modest greenhouse gas reductions, primarily by wealthier industrialized nations. But even this limited agreement faces tough sledding in the ratification process, particularly in the US Senate, where a majority of members have gone on record opposing the Kyoto Protocol. Failure to ratify the agreement in the United States, the world's largest producer of greenhouse gases, could undermine international support for the treaty. Environmentalists, proposing sharp emission reductions in industrialized nations, battled the fossil fuel industry, which sought minimal reductions and the full participation of developing nations. Industrialized nations disagreed among themselves, as well, with large greenhouse gas producers, such as the United States and Japan, fighting for limited immediate emission reductions, and Europe supporting deep cuts. Finally, developing countries, which have blamed industrialized nations for causing most of the emissions, demanded that steep emission, reductions be imposed on the developed world while largely exempting themselves – a position supported by Europe but strongly opposed by the United States.

The Kyoto Protocol calls for 38 industrialized nations to reduce their emissions by an average of 5.2 per cent below 1990 levels by 2012. The United States, which currently emits more than 20 per cent of the world's carbon dioxide from fossil fuels, agreed to a 7 per cent cut below 1990 levels, while 15 nations of the EU committed themselves to an 8 per cent reduction. Japan agreed to a 6 per cent cut. The pact does not require any binding emission reductions for developing countries, but subsequent rounds of negotiations were set up to consider this and other issues in the years ahead.

According to the UN-sponsored Intergovernmental Panel on Climate Change (IPCC), somewhere between 50 per cent and 75 per cent of projected future emissions produced by burning fossil fuels must be eliminated by 2050 to prevent a doubling or even tripling of atmospheric greenhouse gases late in the next century. This means that research, development and deployment programmes for new, clean energy technologies must be established in the coming decades.

Copenhagen Climate Convention

It was held in December 2009. Many issues were raised to control carbon emissions but the developed countries did not cooperate for any resolution. Consequently, this meet was finished without any solution.

To conclude, it can be stated that the global environmental collapse is not inevitable. But, the developed world must work with the developing world to ensure that new industrialized economies do not add to the world's environmental problems. Politicians must think of sustainable development rather than economic expansion. Conservation strategies have to become more widely accepted, and people must learn that energy use can be dramatically diminished without sacrificing comfort. In short, with the technology that currently exists, the years of global environmental mistreatment can begin to be reversed.

CHAPTER 15

India's Environmental Regulations

Environmental conservation has become the prime concern of the governments the world over. Globally, both the developed and developing countries are trying to minimize the environmental degradation through framing and implementing environmental regulations. In the beginning of the 1970s, worldwide discussion on the environment gained momentum partly due to high level of environmental pollution and consequent degradation and partly because of increasing awareness towards conserving the environment. Various regulations on environmental protection were formulated along with forests and wildlife conservation acts. In India, regulations on conserving environment started implementing in the 1970s although, the text on environmental conservation is widely mentioned in the Constitution of India.

The state's responsibility with regard to environmental protection has been laid down under Article 48–A of our Constitution, which states that the state shall endeavour to protect and improve the environment and safeguard the forests and wildlife of the country. Environmental protection is also a fundamental duty of every citizen of this country under Article 51–A (g) of our Constitution as it says, "It shall be the duty of every citizen of India to protect and improve the natural environment including forests, lakes, rivers and wildlife and to have compassion for living creatures." While Article 48–A comes under Directive Principles of State Policy, Article 51–A (g) comes under Fundamental Duties. The state's responsibility with regard to raising the level of nutrition and the standard of living and to improve public health has been laid down under Article 47 which says, "The State shall regard the raising of the level of nutrition and the standard of living of its people and the improvement of public health as among its primary duties and, in particular, the State shall endeavour to bring about prohibition of the consumption except for medicinal purposes of intoxicating drinks and of drugs which are injurious to health." The 42nd amendment to the Constitution, which was brought about in the year 1974, makes it the responsibility of the State Government to protect and improve the environment and to safeguard the forests and wildlife of the country. The latter, under Fundamental Duties, makes it the fundamental duty of every citizen to protect and improve the natural environment including forests, lakes, rivers and wildlife and to have compassion for living creatures. The subjects related to environment in the seventh schedule of the Constitution in three lists are given in Box 15.1.

Box 15.1 Environmental Regulations in the Constitution

I. Union List

- Industries.
- Regulation and development of oil fields and mineral oil resources.
- Regulation of mines and mineral development.
- Regulation and development of inter-state rivers and river valleys.
- Fishing and fisheries beyond territorial waters.

II. State List

- Public health and sanitation.
- Agriculture, protection against pest and prevention of plant diseases.
- Land, colonisation, etc.
- Fisheries.

Cont'd...

...Cont'd

- Regulation of mines and mineral development subject to the provisions of List I.
- Industries subject to the provisions of List I.

III. Concurrent List

- Forests.
- Protection of wild animals and birds.
- Economic and social planning.
- Population control and family planning.

As conferred by Article 246(1), while the Union Government is supreme to make any law over the subjects enumerated in List I, the states, under Article 246 (3), enjoy competence to legislate on the entries contained in List II, and both the Union and the states under Article 246(2) have concurrent jurisdiction on entries contained in List III. In the event of a clash, the Union enjoys a primacy over states in that its legislation in the Union and the Concurrent Lists prevails over state legislations. Also, the parliament has residuary powers to legislate on any matter not covered in the three lists (Article 248). The major regulations related to the environment are discussed in the following paragraphs:

Wildlife Protection Act, 1972

The wildlife laws have a long history and are the result of an increasing awareness of the compelling need to restore the catastrophic ecological imbalances introduced by the depredations inflicted on nature by human being. The earliest codified law can be traced to 3rd century B.C., when Ashoka, the king of Maghadha, enacted a law for preservation of wildlife and environment. But, the first codified law in India, which heralded the era of laws for the wildlife and protection, was enacted in the year 1887 by the British and was titled as the Wild Birds Protection Act, 1887. This Act enabled the then government to frame rules prohibiting the possession or sale of any kinds of specified wild birds, which have been killed or taken during the breeding season. Again, the colonial government in the year 1912 passed the Wild Birds and Animals Protection Act, 1912 as the Act of 1887 proved to be inadequate for the protection of wild birds and animals. The Act of 1912 was amended in the year 1935 by the Wild Birds and Animals Protection (Amendment) Act, 1935. After the Second World War, the freedom struggle for India started taking its shape and wildlife was relegated to the background. But after independence, the Constituent Assembly in the Draft Constitution placed "Protection of Wild Birds and Wild Animals" at entry 20 in the State List and the State Legislatures have been given power to legislate on this. It was not till late 1960s that the concern for the depletion of wildlife finally aroused.

According to the Wildlife Protection Act, 1972, "wildlife" includes any animal, bees, butterflies, crustacean, fish and moths; and aquatic or land vegetation which forms part of any habitat. In accordance with Wildlife (Protection) Amendment Act, 2002, no alternation of boundaries/national park/sanctuary shall be made by the state government except on recommendation of the National Board for Wildlife (NBWL). Further, in terms of Supreme Court order of 13 November 2000, the state governments have to seek prior permission of Supreme Court before submitting the proposal for diversion of forestland in national parks/sanctuaries. Whenever, any part of wildlife sanctuary/national park is getting affected by a hydroproject, the forest proposal in respect of such project is entertained by the Ministry of Environment and Forests (MoEF), Government of India only after permission of de-reservation/de-notification of wildlife sanctuary/national park has been accorded. After recommendation of Standing Committee of NBWL proposal for de-reservation/de-notification can be ratified by the Supreme Court.

Water Prevention and Control Act, 1974

The objectives of the Water Prevention and Control of Pollution Act, 1974 are to provide for the prevention and control of water pollution and the maintenance or restoration of the wholesomeness of water for the establishment. With a view to carrying out the aforesaid purposes, the prevention and control of water pollution, assign powers and functions through this act.

The Forest Conservation Act, 1980

This Act refers to the conservation of forests and regulating diversion of forestlands for non-forestry purposes. When projects fall within forestlands, prior clearance is required from relevant authorities under the Forest Conservation Act, 1980. State governments cannot de-reserve any forestland or authorize its use for any non-forest purposes prior without prior approval from the Central government. The steps for forest clearance are briefly described below:

Identification of Forest Area Involved

Preliminary location of project is done by using tools such as the forest atlas and Survey of India maps. During route alignment, all possible efforts are made to avoid the forest area (like national park and sanctuaries) or to keep it to the barest minimum. Whenever it becomes unavoidable due to the geography of terrain or heavy cost involved in avoiding it, different alternative options are considered to minimize the requirement of forest area. For selection of optimum proposal, the following criteria are taken into consideration:

- any monument of cultural or historical importance is not affected by the project;
- the proposed alignment of the project line does not create any threat to the survival of any community with special reference to tribal community;
- the proposed alignment of the project does not affect any public utility services such as playgrounds, schools and other establishments;
- the alignment of the project does not pass through any sanctuaries, national park, biosphere reserves or eco-sensitive zones; and
- the alignment of the project does not infringe with area of natural resources.

To achieve this, selection of forest area involved is undertaken in close consultation with representatives from the state forest and revenue departments. Minor alterations are made to avoid environmentally-sensitive areas and settlements at execution stage. Trees on such locations are felled but after stringing is complete and natural regeneration is allowed to specific heights and whenever required the tree plantation is taken.

Formulation of Forest Proposal

After finalization of forest area involved for project location Satluj Jal Vidyut Nigam, Ltd. (SJVNL) submits details in prescribed proforma to the respective Divisional Forest Officer (DFO)/Nodal Officer (Forest) of concerned state government, who forwards it to the concerned DFO/Conservator of Forest for processing of clearance under the Forest (Conservation) Act, 1980. The DFO then surveys the relevant forest area required for the construction of project under the possible alternatives. Forest authorities also conduct a cost-benefit analysis to assess the loss of forest produce, loss to environment vis-à-vis benefits of project. Compensatory afforestation scheme is prepared to compensate loss of vegetation and is the most important and integral part of the proposal. For compensatory afforestation, the forest authorities identify degraded forestland of twice the area of affected land. SJVNL provides undertaking/certificate to meet the cost of compensatory afforestation and the net present value (NPV) of forestland diverted. The NPV rate varies from Rs 5.8 to Rs 9.2 lakh per hectare (as per MoEF notification of 23 April 2004 and is payable to the Compensatory Afforestation Fund Management and Planning Authority (CAMPA). If the forest is rich in wildlife, then the Chief Wildlife Warden (CWW) also gets a detailed assessment report prepared including measures to protect the wildlife, which is submitted with the proposal.

Approval of Proposal

The proposal is submitted to the state forest department and then forwarded to the Principal Chief Conservator of Forests in the state and finally to the state government. The state government recommends the proposal for further processing and approval to (a) concerned regional office of the MoEF if the area involved is 40 hectare or less or (b) MoEF, New Delhi if the area is more than 40 hectare. To facilitate speedy approval of the proposal involving lesser area, the MoEF had established regional offices in each region for processing and approving these proposals. The MoEF approves the proposal in two stages. In first stage it is accorded with certain conditions depending upon the case. Second stage or final approval is accorded after the compliance report of the conditions stipulated in first forest proposal is received by MoEF, from State Forest Department. SJVNL follows all relevant guidelines including the directions of the Supreme Court in this regard from time to time.

Air Prevention and Control of Pollution Act, 1981

The objective of this Act is to provide for the prevention, control and abatement of air pollution. At the United Nations Conference on the Human Environment held at Stockholm in June 1972, in which India also participated, it was decided by the participating nations to take appropriate steps for the preservation of the natural resources of the earth which, among other things, includes the preservation of the quality of air and control of air pollution. Therefore, it is considered necessary to implement the decisions in so far as they relate to the preservation of the quality of air and control of air pollution.

Environmental Protection Act, 1986

The Environment Protection Act, 1986 was introduced as an umbrella legislation that provides a holistic framework for the protection and improvement of the environment. In terms of responsibilities, the Act and the associated rules require for obtaining environmental clearances for specific types of new/expansion projects (addressed under Environmental Impact Assessment Notification, 1994) and for submission of an environmental statement to the State Pollution Control Board annually. Environmental clearance is not applicable to hydroprojects also. SJVNL undertakes Environmental Impact Assessment (EIA) for all projects as a standard management procedure as laid down in the, 1986 Act and also functions within permissible standards of ambient air quality and noise levels as prescribed by national laws and international regulations.

Ozone Depleting Substances and Control Rules, 2000

The MoEF, vide its notification dated 17th July, 2000 under the sections 6, 8 and 25 of the Environment (Protection) Act, 1986 has notified rules for regulation/control of ozone depleting substances (ODS) under Montreal Protocol. As per the notification certain controls and regulations have been imposed on manufacturing, import, export and use of these compounds. Organizations as per provisions of notification shall phase out all equipments, which use these substances, in order to be CFC-free organization in near future.

The Biological Diversity Act, 2002

The MoEF has enacted the Biological Diversity Act, 2002 under the United Nations Convention on Biological Diversity signed at Rio de Janeiro on 5th June, 1992 to which India is also a party. This Act is aimed to provide for the conservation of biological diversity, sustainable use of its components, and fair and equitable sharing of the benefits arising out of the sued of biological resources, knowledge and for matters connected therewith or incidental thereto. As per the provision of Act, certain areas, which are rich in biodiversity and encompass unique and representative ecosystems, are identified and designated as 'biosphere reserve' to facilitate its conservation. All restrictions applicable to protected areas like national parks and sanctuaries are also applicable to these reserves. SJVNL abides by the provisions of Act wherever applicable and try avoiding these biosphere reserves while finalizing the project infrastructure locations.

The Electricity Act, 2003

This Act seeks to create a framework for the power sector development by measures conducive to the industry. It, however, does not explicitly deal with environmental implications of activities related to power transmission. Under the Section 68(1) of the Act sanction from the Ministry of Power is a mandatory requirement for taking up any new project. The sanction authorizes SJVNL to plan and coordinate activities to commission new projects.

Hazardous Waste Management and Handling Amendment Rules, 2003

These rules classify used mineral oil as hazardous waste that requires its proper handling and disposal. Organizations will seek authorization for disposal of hazardous wastes from concerned State Pollution Control Boards as and when required.

Other Environment-Related Laws

- The Shore Nuisance (Bombay and Kolaba) Act, 1853: This is the earliest Act on the statue book concerning control of water pollution in India.
- The Serais Act, 1867: The Act enjoined upon a keeper of Serai or an inn to keep a certain quality of water fit for consumption by "persons and animals using it" to the satisfaction of the District Magistrate or his nominees. Failure for maintaining the standard entailed a liability of Rs 20.
- The North India Canal and Drainage Act, 1873: Certain offences have been listed under the Act contained in Section 70.
- Obstruction in Fairways Act, 1881: Section 8 of the Act empowered the Central Government to make rules to regulate or prohibit the throwing of rubbish in any fairway leading to a port causing or likely to give rise to a bank or shoal.
- Indian Easements Act, 1882: Illustrations (f), (h) and (j) of Section 7 of the Act deal with pollution of water.
- The Indian Fisheries Act, 1897: The Indian Fisheries Act, 1897 contains seven sections. Section 5 of the Act prohibits destruction of fish by poisoning waters.
- Indian Ports Act, 1908: Water pollution by oil has been regulated by this Act.
- The Indian Forest Act, 1927: Section 26(i) of the Act makes it punishable if any person, who, in contravention of the rules made by the state government, poisons water of a forest area. The state government has been empowered under Section 32(f) to make rules relating to poisoning of water in forests.
- The Damodar Valley Corporation Act, 1948: The Act relates to make regulations with the previous sanction of the central government for preventing pollution of water.
- The Factories Act, 1948: It is a sort of social welfare legislation intended to secure health, safety and welfare of the workers employed in factories. However, some of the provisions of this Act are concerned with the prevention of water pollution.

- The Mines Act, 1952: Chapter V of the Act deals with provisions regarding health and safety of the employees. Section 19(i) entrusts the government responsibility for arrangement for the quality of water for drinking purposes.
- The River Boards Act, 1956: The Act provides for the creation of River Boards for regulation and development of inter-state rivers and river valleys. One of the functions of the Board is to advise to the government concerned on prevention of pollution of the waters of the inter-state rivers.
- The Merchant Shipping Act, 1958: The International Convention for the Prevention of Pollution of the Sea by Oil, 1954 is the first treaty for the reduction of oil pollution of the sea. In order to give effect to this Convention, the Merchant Shipping Act regulates and controls the discharge of oil or oil mixture by an Indian tanker or ship within any of the prohibited zones or by a foreign tanker or other ship within the prohibited zone adjoining the territories of India. Further, there is a prohibition for discharging any oil anywhere at sea from an Indian ship.

Box 15.2 Legal and Regulatory Framework for Hydropower Projects

The principal environmental regulatory agency in India is the MoEF. It formulates environmental policies and accords environmental clearance for the projects. The State Pollution Control Board (SPCB) accords 'No Objection Certificate' (NOC) and 'Consent for Establishment and Operation' for the projects. The project features entail an EIA Study to be conducted which is a prerequisite for obtaining environmental clearance from MoEF. EIA is an important management tool for ensuring optimal use of natural resources for sustainable development, and was introduced in India initially for River Valley Projects in 1978–79. The scope of the EIA has been enhanced to cover other developmental sectors such as industries, mining schemes, energy, etc. To facilitate project proponents in collection of environmental data and formulation of environmental management plans, it is now mandatory

Cont'd...

...Cont'd

under the Environment (Protection) Act, 1986, for different categories of developmental activities involving investment beyond certain thresholds. The notification was issued on 27th January 1994 and was amended on 4th May 1994. Thus, it is hoped that it would provide an opportunity both for the project proponents and government to assess the impact of the concerned project on the environment before it actually comes into play. The EIA study document fulfils the requirements for environmental clearance from various agencies at the state level. These include SPCB and Committee of Experts working under the aegis of Department of Science and Technology of State Governments.

In India, after independence, many environmental regulations were framed and implemented from time to time to conserve natural resources and the environment at a large scale. In the Constitution of India, regulation regarding conservation of the environment is given in central, state and concurrent lists. But, the awareness towards environmental conservation was truly started in the 1970s and thereafter vigorous terms and conditions were imposed. It had positive impacts on many of the states, particularly in the Indian Himalayan states where the state governments took several other initiatives for environmental conservation. Still in many areas of the countries these rules are being violated and due to that large-scale illegal felling of trees are continued. Framing the regulations are always easy but their implementation remains inactive throughout the history of conservation. Indeed, if all regulations, which made for conserving environment, are implemented in a true sense, the problems of environmental degradation will be solved largely.

Appendices

List of Applicable Indian Legislations

Air (Prevention and Control of Pollution) Act, 1981 (as amended in 1987)

Air (Prevention and Control of Pollution) Rules, 1983

Ancient Monuments and Archaeological sites and Remains Act, 1958

Ancient and Historical monuments and Archaeological Sites and Remains Rules, 1971

Building and Other Construction Workers Act, 1996

Bureau of Indian Standards on safety and health

Coastal Regulation Zone (CRZ) Notification, December 1990 (as amended in June 1997)

Constitution of India, 42nd Amendment, Articles 48A and 51A(g)

Dangerous Machines (Regulation) Act, 1983

Dock Workers (Safety, Health and Welfare) Act, 1986

Dock Workers (Safety, Health and Welfare) Rules, 1989

Emergency Planning, Preparedness and Response for Chemical Accident Rules, 1996

Environment (Protection) Act, 1986

Environment (Protection) (Amendment) Rules, 1996

Environmental Impact Assessment Notification, January 1994 (as amended in May 1994)

Environmental Standards Notification for specific industrial and other sources, December 1998

Factories Act, 1948

Factories Rules

Factories (Amendment) Act, 1987

Factories (Amendment) Act, 1987 – Model Rules

Fly Ash Utilization Draft Notification, May 1998

Forest Conservation Act, 1927

Forest Conservation Act 1980 (as amended in 1988)

Forest Conservation Rules 1981 (as amended in 1992)

Gas Cylinder Rules, 1981

Hazardous Waste Management Rules, 1989

Hazardous Wastes (Management and Handling) Amendment Draft Rules, 1999

Indian Boilers Act, 1923

Indian Electricity Act, 1910

Indian Electricity Rules, 1956

Indian Explosives Act, 1884

Indian Explosive Rules, 1983

Indian Roads Congress, IRC 104:1988 – Guidelines for Environmental Assessment of Highway Projects

Insecticides Act, 1968 and the Insecticide Rules, 1971

Land Acquisition Act, 1989

Land Compensation Act, 1894 (as amended in 1984)

Manufacture, Storage and Import of Hazardous Chemical (MSIHC) Rules, 1989

Manufacture, Storage and Import of Hazardous Chemical (Amendment) Draft Rules, 1999

Merchant Shipping Act, 1954 and Merchant Shipping Amendment Act, 1970

MoEF's Consolidated Guidelines for Diversion of Forest Land under the Forest (Conservation) Act, 1980

MoEF's EIA Questionnaires, 1999

MoEF's Environmental Guidelines for Rail/Road/Highway Projects (1986–1989)

MoEF's Handbook of Environmental Procedures and Guidelines, 1994

Motor Vehicles Act, 1988

Motor Vehicles (Central) Rules, 1989

National Environment Appellate Authority Act, 1997

Petroleum Act, 1934

Petroleum Rules, 1976

Public Hearing Notification, April 1997

Public Liability Insurance Act

Rehabilitation and Resettlement (Central and State Legislation)

Static and Mobile Pressure Vessels (Unfired) Rules, 1981

Thermal Power Plants Notification, April 1997

Thermal Power Plants Notification on Ash Content of Coal, Sept. 1997

Town and Country Planning Act 1971

Water Cess (Prevention and Control of Pollution) Act, 19?? (as amended 1988)

Water Cess (Prevention and Control of Pollution) Rules, 19??

Water (Prevention and Control of Pollution) Act, 1974 (as amended 1988)

Water (Prevention and Control of Pollution) Rules, 1983

Wild Birds and Animals Act 1912

Wildlife (Protection) Act, 1972

Environmental Calendar of Activities

There are several days of special environmental significance which can be celebrated in the community and can be used for creating environmental awareness.

February 2: World Wetland Day is celebrated to create awareness about wetlands and their value to mankind. On February 2nd 1971, the

Ramsar Convention on Wetlands of International importance was signed at Ramsar in Iran. You can initiate a campaign for proper use and maintenance of wetlands in the vicinity of the city or village.

March 21: World Forestry Day can be used to initiate a public awareness campaign about the extremely rapid disappearance of our forests. The programme must be action oriented and become an ongoing process with activities such as tree plantation.

April 7: World Health Day – The World Health Organization (WHO) came into existence on this day in 1948. A campaign for personal sanitation and hygiene for understanding issues of public health, occupational health, etc., can be carried out. Topics that deal with environment related diseases and their spread can be discussed and preventive measures suggested.

April 18: World Heritage Day can be used to arrange a visit to a local fort or museum. Environment also includes our cultural monuments. Students could use this opportunity to create awareness among the local people about their very valuable heritage sites.

April 22: Earth Day was first celebrated in 1970 by a group of people in the USA to draw attention to increasing environmental problems caused by humans on earth. This day is now celebrated all over the world with rallies, festivals, clean-ups, special shows and lectures.

June 5: World Environment Day marks the anniversary of the Stockholm Conference on Human Environment in Sweden in 1972, where nations of the world gathered to share their concern over human progress at the expense of the environment. This day can be used to project the various environmental activities that the college has undertaken during the year. New pledges must be made to strengthen an environmental movement at the college level.

June 11: World Population Day is a day when the vital link between population and environment could be discussed in seminars held at college and other NGOs.

August 6: Hiroshima Day could be used to discuss our own Bhopal Gas Tragedy and the Chernobyl disaster.

September 16: World Ozone Day was proclaimed by the United Nations as the International Day for the preservation of the ozone layer. This is a good occasion for students to find out more about the threats to this layer and initiate discussion on what they can do to help mitigate this global threat. The day marks the Montreal Protocol signed in 1987 to control production and consumption of ozone depleting substances.

September 28: Green Consumer Day could be used to create an awareness in consumers about various products. Students could talk to shopkeepers and consumers about excess packaging and a campaign to use articles which are not heavily packaged could be carried out.

October 1–7: Wildlife Week can consist of seminars on conserving our species and threatened ecosystems. The State forest Departments organize various activities in which every student should take part. A poster display, a street play to highlight India's rich biodiversity can be planned. Wildlife does not only mean animals, but includes plants as well.

Do's and Don'ts

Most of us are always complaining about the deteriorating environmental situation in our country. We also blame the government for inaction. However how many of us actually do anything about our own environment? You can think about the things you can do that support the environment in your daily life, in your profession and in your community. You can make others follow your environment friendly actions. A famous dictum is to 'think globally and act locally' to improve your own environment. 'You' can make a difference to our world.

Biodiversity Conservation: A great proportion of the residual wilderness of India is now under great threat. Its unique landscapes are shrinking as the intensive forms of agriculture and industrial growth spreads through a process called 'development'. Modern science has serious doubts about the possibility of the long-term survival of the human race if man continues to degrade natural habitats, extinguishes millions of years of evolution through an extinction spasm, and looks only at short-term gains. The extinction of species cannot be reversed. Once a species is lost, it is gone forever. Future generations will hold us responsible for this great loss. We frequently forget that we are a part of a great complex web of life and our existence depends on the integrity of 1.8 million species of plants and animals on earth that live in a large number of ecosystems. The following are some of the things you can do to contribute towards our ecological security and biodiversity conservation.

Do's

1. Plant more trees of local or indigenous species around your home and your workplace. Encourage your friends to do so. Plants are vital to our survival in many ways.

2. If your urban garden is too small for trees, plant local shrubs and creepers instead. These support bird and insect life that form a vital component of the food chains in nature. Urban biodiversity conservation is feasible and can support a limited but valuable diversity of life.
3. If you live in an apartment, grow a terrace or balcony garden using potted plants. Window boxes can be used to grow small flowering plants, which also add to the beauty of your house.
4. Whenever and wherever possible prevent trees from being cut, or if it is not possible for you to prevent this, report it immediately to the concerned authorities. Old trees are especially important.
5. Insist on keeping our hills free of settlements or similar encroachments. Degradation of hill slopes leads to severe environmental problems.
6. When shopping, choose products in limited packaging. It will not only help cut down on the amount of waste in landfills, but also helps reduce our need to cut trees for paper and packaging.
7. Look for ways to reduce the use of paper. Use both sides of every sheet of paper. Send your waste paper for recycling.
8. Buy recycled paper products for your home. For example sheets of paper, envelopes, etc.
9. Reuse cartons and gift-wrapping paper. Recycle newspaper and waste paper instead of throwing it away as garbage.
10. Donate used books and magazines to schools, hospitals, or libraries. The donations will not only help these organizations, but also will reduce the exploitation of natural resources used to produce paper.
11. Participate in the events that highlight the need for creating Sanctuaries and National Parks, nature trails, open spaces, and saving forests.
12. Support Project Tiger, Project Elephant, etc., and join NGOs that deal with environmental protection and nature conservation.
13. Involve yourself and friends in activities carried out during Wildlife Week and other public functions such as tree plantation drives and protests against destruction of the environment.

Don'ts

1. Do not present flower bouquets instead give a potted plant and encourage your friends to do so.

2. Do not collect unnecessary pamphlets and leaflets just because they are free.
3. Do not use paper plates and tissues or paper decorations when you hold a party.

Habitat Preservation: The rapid destruction of forests and the growth of human habitations and activities have reduced the natural habitats of animals and birds. Loss of habitat is one of the major pressures on several species and has led to the extinction of several rare and endemic species. Many others are seriously threatened. We therefore, have the responsibility to preserve remaining habitats and their inhabitants. The following are some 'dos and don'ts' that can help preserve threatened ecosystems.

Do's

1. Visit forests responsibly. Remember to bring out everything you take in, and clean up litter left by others. Stay on marked trails, and respect the fact that wildlife need peace and quiet. Study the ecosystem; it gives one a greater sense of responsibility to conserve it.
2. Be kind to animals. Stop friends from disturbing or being cruel to wild creatures such as birds, frogs, snakes, lizards and insects.
3. Learn about birds. Identify birds that are common in your area. Understand their food requirements and feeding habits. Construct artificial nesting boxes for birds. This will encourage birds to stay in your neighbourhood, even if their nesting habitat is scarce. You can learn more about birds by making a birdbath. Birds need water to drink and to keep their feathers clean. You can make a birdbath out of a big ceramic or plastic saucer. Having birds around your home, school or college can even help increase species diversity in the area.
4. Attract wildlife such as small mammals, such as squirrels, to your garden by providing running or dripping water. Make a hole in the bottom of a bucket and poke a string through to serve as a wick. Hang a bucket on a tree branch above your birdbath to fill it gradually with water throughout the day.
5. Protect wildlife, especially birds and insects that are insectivorous and live in your neighbourhood by eliminating the use of

chemicals in your garden. Instead, use organic measures from vermicomposting and by introducing natural pest predators. Do your gardening and landscaping using local plants, to control pests in your garden. If you have pets, feed them well and give them a proper home and in an emergency proper medical care.

6. When you visit a zoo learn about the animals that are found there but do not tease or hurt them through the bars of their cage. They have a right to a peaceful existence. The zoo is in any is case not an ideal home for them
7. Do not buy products like purses, wallets, boots and that are made from reptile skins. If you are not certain that a product is made from a wild species, its better to avoid using it.
8. Do not buy products made from ivory. Elephants are killed for their tusks, which are used to make a variety of ivory products.
9. Do not use any wild animal or plant products that are collected from the wild and have dubious medicinal properties. You may be endangering a species and even your own health.

Don'ts

1. Do not disturb, tease, hurt or throw stones at animals in a Protected Area and stop others from doing so. If you see an injured animal contact the Forest Officials.
2. Do not disturb or destroy the natural habitats of birds or animals.
3. Do not use articles like leather handbags and lipsticks, which are made from animal products. No wildlife products should be used.
4. Do not catch or kill butterflies or other insects. Butterflies, moths, bees, beetles and ants are important pollinators.
5. Do not kill small animals and insects like dragonflies and spiders, as they act as biological pest control mechanisms.
6. Do not bring home animals or plants collected in the wild. You could be seriously harming wild populations and natural ecosystems where they were collected.

Soil Conservation: Soil degradation affects us all in some way, either directly or indirectly. There are many ways that each of us can help in solving environmental problems due to loss of soil. Following are some of the dos and don'ts for conserving soil.

Do's

1. Cover the soil in your farm or garden with a layer of mulch to prevent soil erosion in the rains and to conserve soil moisture. Mulch can be made from grass clippings or leaf litter.
2. If you plan to plant on a steep slope in your farm or garden, prevent soil erosion by first terracing the area. Terraces help in slowing the rain water running downhill so it can soak into the soil rather than carry the soil away.
3. Help prevent soil erosion in your community by planting trees and ground-covering plants that help hold the soil in place. You might organize a group of citizens to identify places that need planting, raise funds, work with the local government to plant trees, shrubs and grasses, and maintain them over the long term.
4. If your college is surrounded by open space, evaluate how well the soil is being conserved. Look for places where soil can run off, like on an unplanted steep slope or stream bank, or where soil is exposed rather than covered with mulch. These areas need special care and must be carefully replanted.
5. Add organic matter to enrich your garden soil. For example, compost from kitchen scraps and manure from poultry, cows are good sources of nutrients. Make sure manure is not too fresh and that you do not use too much. Healthy soil grows healthy plants, and it lessens the need for insecticides and herbicides.
6. In your vegetable garden, rotate crops to prevent the depletion of nutrients. Legumes such as peas and beans put nitrogen back into the soil.
7. Set up a compost pit in your college or garden, so that you can enrich your soil with the organic waste from the kitchen and cut down on the amount of waste it sends to a landfill. Set up buckets in your college or lunchroom where fruit and left-over food can be put. Empty the buckets daily into a compost pit, and use the rich compost formed in a few weeks to enrich the soil around the college. Kitchen scraps, leaves and grass clippings are excellent compost.
8. Encourage your local zoo, farms, and other organizations or people that house a large number of animals to provide your community with bioferilizer made from animal manure. This can be composted to make a rich fertilizer, and it forms an additional source of income for the animal owners.

9. Buy organically grown produce to help reduce the amount of toxic pesticides used in farms that harm soil organisms. Look for organically grown produce in your grocery shop, or try growing some yourself if you have the space.
10. Support environmental campaigns in your State and community. Cutting down on irresponsible development can protect soil, biodiversity, and enhance our quality of life.

Don'ts

1. Do not remove grass, leave it on the lawn. Cuttings serve as moisture-retention mulch and a natural fertilizer.
2. Do not use toxic pesticides in your garden-they often kill the beneficial organisms, your soil needs to stay healthy.

Conserving Water: Most of India has good average annual rainfall, however we still face a water shortage nearly everywhere. This is one of the major environmental problems in our country. Conservation of this very precious natural resource is very important and it is the need of the hour. It should start with every individual. It must start with you! Following are some of the things you can do to conserve this precious natural resource.

Do's

1. Reduce tthe amount of water used for daily activities. For example – turn off the tap while brushing your teeth to save water.
2. Reuse the rinsing water for house-plants. Reuse the water that vegetables are washed in to water the plants in your garden or your potted plants.
3. Always water the plants early in the morning to minimize evaporation.
4. Soak the dishes before washing them to reduce water and detergent usage.
5. Look for leaks in the toilet and bathroom to save several litres of water a day.
6. While watering plants, water only as rapidly as the soil can absorb the water.
7. Use a drip irrigation system to water more efficiently.
8. When you need to drink water, take only as much as you need to avoid wastage. So many people in our country don't even have access o clean drinking water.

9. Saving precious rainwater is very important. Harvest rainwater from rooftops and use it sustainably to recharge wells to reduce the burden on rivers and lakes.
10. Monitor and control wastes going into drains for preventing water pollution.
11. Replace chemicals like phenyl, strong detergents, shampoo, chemical pesticides and fertilizers used in your home, with environment friendly alternatives, such as neem and biofertilisers. Groundwater contamination by household chemicals is a growing concern.
12. For Ganesh Chaturthi, bring home a 'Shadu' idol instead of a Plaster of Paris idol and donate it instead of immersing it in the river to reduce river pollution.

Don'ts

1. Do not turn your tap on full force, instead maintain a slow flow.
2. Do not use a shower; instead use a bucket of water for bathing. A 10 minute shower wastes many litres of water as compared to using water from a bucket.
3. Do not over water garden plants, water them only when necessary.
4. Do not pollute sources of water or water bodies by throwing waste into them. This is the water you or someone else has to drink.
5. Do not throw waste into toilets because finally it goes into water bodies.

Conserving Energy: Coal, petroleum and oil are non-renewable mineral resources of energy. At the current rate of fossil fuel consumption, the present oil reserves on the earth will last only for the next 30 to 50 years. Crores of rupees are being spent to extract process and distribute coal, petroleum and electricity. Experiments are being carried out to generate energy from wind, and photovoltaic cells. They are highly successful. At an individual level, every one of us should try to conserve energy. Following are some of the things you can do to conserve energy.

Do's

1. Turn off the lights, fans and air conditioning when not necessary.
2. Use low voltage lights.
3. Use tube lights and energy saver bulbs as they consume less electricity.

4. Switch off the radio and television when not required.
5. Use alternative sources of energy like solar power for heating water and by cooking food in a solar cooker.
6. Cut down on the use of electrical appliances.
7. In summer, shut windows, curtains and doors early in the morning to keep the house cool.
8. Use a pressure cooker as much as possible to save energy.
9. Turn off the stove immediately after use.
10. Plan and keep things ready before you start cooking.
11. Keep vessels closed while cooking and always use small, narrow mouthed vessels to conserve energy.
12. When the food is almost cooked, switch off the gas stove and keep the vessel closed. It will get completely cooked with the steam already present inside.
13. Soak rice, pulses, etc., before cooking to reduce cooking time and save fuel.
14. Get your family to eat together, it will save re-heating fuel.
15. Select a light shade of paint for walls and ceilings, as it will reflect more light and reduce electrical consumption.
16. Position your reading tables near the window and cut down on your electricity bill by reading in natural light.
17. Use a bicycle it occupies less space, releases no pollutant and provides healthy exercise.
18. Try using public transport systems like trains and buses as far as possible.
19. Plan your trips and routes before setting out.
20. Walk rather than drive wherever possible. Walking is one of the best exercises for your health.
21. Get vehicles serviced regularly to reduce fuel consumption and reduce pollution levels.

Don'ts

1. Do not use unnecessary outdoor decorative lights.
2. Do not use a geyser during summer. Instead, heat water naturally with the help of sunlight.
3. Do not use halogen lamps as they consume a lot of electricity.
4. Do not put food in the refrigerator when they are still hot.

Bibliography

Arvill, R. 1967: *Man and Environment: Crisis and the Strategy of Choice*, Penguin, Harmondsworth.

Atkinson, B.W. 1981: *Precipitation in Man and Environment Processes* (Ed) KJ Gregory and DE Walling, Butterworths, pp. 23–37.

Aultfather, W. and Crozier, E.S. 1971: A Resource Inventory and Planning System for Wildlife Area, *Journal of Wildlife Management*, Vol. 35, pp. 168–74.

Batisse, M. 1986: Developing and Focusing the Biosphere Reserve Concept, *Nature and Resources*, Vol. 22 (3), pp. 2–11.

Beazley, R. 1967: Conservation Decision – Making a Rationalization, *Natural Resources Journal*, Vol. 7, pp. 345–60.

Botkin, D.B. and Keller, EA. 1982: *Environmental Studies*, CE Merrill Publishing Company, Columbus, pp. 505.

Brew, D.A. 1974: *Environmental Impact Analysis: The Example of the Proposed Trans-Alaska Pipeline*, US Geological Circular 695.

Brock, T.D. 1967: The Ecosystem and the Study State, *Biosciences*, Vol.17, pp. 166–9.

Clapham, B.W. 1973: *Natural Ecosystem*. Macmillan, London.

Cox, C.B., Hedley, I.N. and Moose, P.D. 1973: *Biogeography: An Ecological and Evolutionary Approach*, Blackwell, Oxford.

Crossland, J. 1978: Reporting Pollution, *Environment*, Vol. 20. pp. 29–31.

Darlington, P.J. 1957. *Zoogeography*, John Wiley, New York.

——. 1965. *Biogeography of the Southern End of the World*. Harvard University Press, Cambridge.

Dassman, R.F. 1975: *The Conservation Alternative*. Wiley, New York.

——. 1976. *Environmental Conservation*. Wiley, New York.

Detwyler, T.R. 1971: *Man's Impact on Environment*. Mc Graw-Hill, New York.

Ehrenfeld, D.W. 1970: *Biological Conservation*. Holt, Rinehart and Winston, New York.

Elton, CS. 1958. *The Ecology of Invasion by Plants and Animals*, Methuen, London.

Embleton, C. 1989: Natural Hazards and Global Change, *ITC Journal* 1989–3, 4. pp. 169–178.

Encarta, Encyclopedia. 2005. Natural Resources. CD Rom.

Furley, P.A. and Newey, W.W. 1983: *Man and the Biosphere*, Butterworths, London.

Goudie, A. 1984: *The Nature of the Environment*, Bassil Blackwell Publishers Ltd. pp. 331.

Grime, J.P. 1974: *Vegetation Classification by Reference to Strategies*, Nature, Vol. 250, pp. 26–31.

Hill, A.R. 1975: *Biogeography as a Subfield of Geography Area*, Vol. 7. pp. 156–60.

Kumra, V.K. 1982: *Kanpur City: A Study in Environmental Pollution*, Tara Book Agency, Varanasi, pp. 205.

Longman, K.A. and Jenik, J. 1974: *Tropical Forest and Its Environment*. Longman, London.

Nicholson, M. 1972: *The Environmental Revolution*. Penguin, Harmondsworth.

Nimpuno, K. 1989: Disasters and Social Response, *ITC Journal*, 1989–3, 4. pp. 179–182.

Odum, EP: 1959: *Fundamentals of Ecology*. Saunders, Philadelphia.

Park, C.C. 1980: *Ecology and Environmental Management*, Butterworths, London.

Sati, V.P. 2010: Climate Change Conventions: From Kyoto to Copenhagen, *Pratiyogita Darpan* (English). January 2010.

——. 2007: Environmental Hazards: Type, Distribution, and Mitigation. *Pratiyogita Darpan* (English). April 2007.

——. 2007: Environmental Hazards: Type, Distribution, and Mitigation. *Pratiyogita Darpan* (English). March 2007.

——. 2006: Economic Growth and Environmental Conservation. *Pratiyogita Darpan* (English) November 2006.

——. 2006: Population Explosion: Causes and Consequences. *Pratiyogita Darpan* (English) August 2006.

Simmons, I.G. 1974: *The Ecology of Natural Resources*. Edward Arnold, London.

Singh, Savindra. 1999: *Environmental Geography*. Prayag Pustak Bhawan, Allahabad. pp. 517.

Whittakker, RH. 1975. *Communities and Ecosystems*. Collier-MacMillan London.

Wikipedia: Internet Version.

Index